Proceedings in Life Sciences

International Congress of Entomology, 15th, Washington, D.C., 1976.

The Role of Arthropods in Forest Ecosystems

Edited by W. J. Mattson

With Contributions by

G. D. Amman   J. T. Callahan   B. W. Cornaby
K. Cromack, Jr.   D. A. Crossley, Jr.
D. L. Dindal   W. M. Fender   R. Fogel
B. Heinrich   D. Inouye   J. C. M. Jonkman
J. F. McBrayer   L. J. Metz   P. A. Morrow
P. Sollins   A. W. Todd   R. L. Todd   D. P. Webb

With 28 Figures

Springer-Verlag
New York   Heidelberg   Berlin

WILLIAM J. MATTSON
U.S. Department of Agriculture, Forest Service
North Central Forest Experiment Station
1992 Folwell Avenue
St. Paul, Minnesota 55108/USA

ISBN 0-387-08296-4 Springer-Verlag New York   Heidelberg   Berlin
ISBN 3-540-08296-4 Springer-Verlag Berlin   Heidelberg   New York

Library of Congress Cataloging in Publication Data. International Congress of
Entomology, 15th, Washington, D.C., 1976. The role of arthropods in forest ecosystems. (Proceedings in life sciences). Includes index. 1. Forest insects—Congresses.
2. Arthropoda—Congresses. 3. Forest ecology—Congresses. I. Mattson, William J.,
1943-   II. Title.
QL461.I72   1976   595'.2'045264   77-23558

Preface

The role of arthropods in forest ecosystems is poorly understood. Yet such knowledge may be critical in order to explain fully the fundamental forces that shape the structure and regulate the functioning of such ecosystems. There are numerous hypotheses about the roles of various arthropods, but few, if any, of these hypotheses have been rigorously tested. Some, however, have been repeated so often and so widely that they are now accepted by many as unequivocal fact. Nothing could be further from the truth.

Forest arthropods which derive most of their sustenance from plants are usually specially adapted for feeding in one of three subsystems—the above-ground plant system, the soil-litter system, or the aquatic stream system. Plant-feeding arthropods in the soil-litter and stream systems are primarily saprophous although many consume significant amounts of microorganisms. Research on the role of arthropods in each of these three subsystems has historically been provincial. Until very recently there has been little effort to collate, assimilate, and synthesize the plethora of findings in even one of these systems—much less all three. This Symposium (at the 15th International Congress of Entomology, Washington, D.C. August 19-27, 1976) was organized for the specific purpose of promoting scientific synthesis. It fulfills one of the first requirements in such endeavors; namely, the juxtapositioning of current knowledge and hypotheses so that similarities can be perceived, insights can be derived, and more elaborate conceptual constructs can be built.

These Proceedings have two Divisions—terrestrial phytophagous arthropods and terrestrial saprophagous arthropods. The Division on aquatic arthropods, which was given at the 15th International Congress of Entomology, is unfortunately not included here. The papers present a variety of perspectives on the roles of a variety of arthropods. However, there are some dominant recurring themes—one is the significant impact of arthropods on quantitative and qualitative aspects of system structure. For example: bark beetles affect the frequency distribution of tree size classes and plant succession; phyllophagous and sucking insects can differentially affect the competitive ability of coexisting eucalypt trees, bumblebees and other pollinating organisms can influence plant species composition as well as the evolution of certain qualitative characteristics of plants and plant communities, ants build nests which become ideal sites for the colonization and establishment of certain plants, soil arthropods alter the physicochemical structure of the soil-litter milieu through pelletizing plant debris and converting some refractory substances (often with the aid of microorganisms) into more mobile, elementary forms, soil Collembola are not merely random assemblages of species but interact with one another forming definite associations.

Another obvious theme which reinforces and is a corollary of the first is the sensitivity and responsiveness of arthropods to system structure—broadly speaking, cover and food. For example, it seems that organisms have evolved an "energy fit" to their normal hosts in their usual environment. This fit evolves in relation to the concentration and distribution of energy, nutrients and plant defenses through space and time. Environmental "patchiness" describes such variation. Although the three subsystems are different in many ways, they do have some similar structural features to which arthropod consumers are sensitive. In all three systems it is apparent that there is (1) a scarcity of N and perhaps other elements like P and Na; (2) an abundance of nutritionally impoverished, nearly indigestible substrates like celluloses and lignins; (3) a pervasive presence of secondary plant compounds that are repellent, deterrent or toxic; and (4) an abundance of microorganisms which are intimately associated with arthropod microhabitats, foods, and bodies.

An organism's life history strategy may vary in relation to these as well as other components of system structure. For example, where food substrates are nutritionally pauperate, many consumers rely heavily on microorganisms (as gut symbionts or as food) to enable them to gain adequate nutrition. Secondary compounds usually slow utilization rates of plant tissue by inhibiting microorganisms and arthropod consumers. Arthropods either avoid the compounds by attacking tissues before or after significant concentrations occur or evolve enzymes or other mechanisms to cope with such substances.

The evidence suggests that arthropods, along with microorganisms, significantly interact with system structure; that is, they affect and are affected by it. It is a continuous, dynamic feed-back loop with rates of change usually subtle, but occasionally rapid and sweeping.

St. Paul, July 1977   W. J. MATTSON

Contents

viii

List of Contributors

AMMAN, G. D., Intermountain Forest and Range Experiment Station, Forest Service, U.S.D.A. Ogden, UT 84401, USA

CALLAHAN, J. T., National Science Foundation, Washington, DC 20550, USA

CORNABY, B. W., Battelle Columbus Laboratories, Columbus, OH 43201, USA

CROMACK, K., JR., Department of Forest Science, Oregon State University, Corvallis, OR 97331, USA

CROSSLEY, D. A., JR., Department of Entomology, Institute of Ecology, University of Georgia, Athens, GA 30602, USA

DINDAL, D. L., Department of Forest Zoology SUNY, College of Environmental Science and Forestry, Syracuse, NY 13210, USA

FENDER, W. M., Department of Entomology, Oregon State University, Corvallis, OR 97331, USA

FOGEL, R., Department of Forest Science, Oregon State University, Corvallis, OR 97331, USA

HEINRICH, B., Department of Entomological Sciences, University of California, Berkeley, CA 94720, USA

INOUYE, D., Department of Zoology, University of Maryland, College Park, MD 20740, USA

JONKMAN, J. C. M., Royal Netherlands Embassy, Scientific Office, 4200 Linnean Avenue NW, Washington, DC 20008, USA

McBRAYER, J. F., Environmental Sciences Division, Building 2024, Oak Ridge National Lab., P. O. Box X, Oak Ridge, TN 37830, USA

METZ, L. J., Southeastern Forest Experiment Station, Forest Service, U.S.D.A., Research Triangle Park, NC 27709, USA

MORROW, P. A., Department of Ecology and Behavioral Biology, University of Minnesota, Minneapolis, MN 55455, USA

SOLLINS, P., Department of Forest Science, Oregon State University, Corvallis, OR 97331, USA

TODD, A. W., Department of Botany and Plant Pathology, Oregon State University, Corvallis, OR 97331, USA

TODD, R. L., Institute of Ecology, Department of Agronomy, University of Georgia, Athens, GA 30602 USA

WEBB, D. P., Department of Entomology, Texas A & M University, College Station, TX 77843, USA

Terrestrial Phytophagous Arthropods

Chapter 1

The Role of the Mountain Pine Beetle in Lodgepole Pine Ecosystems: Impact on Succession

G. D. AMMAN

Introduction

The mountain pine beetle, *Dendroctonus ponderosae* (Coleoptera: Scolytidae), is the most aggressive member of its genus in the western United States. Populations of the beetle periodically build up and kill most of the large dominant lodgepole pines, *Pinus contorta* var. *latifolia,* over vast acreages. The beetle is indigenous to North America and probably has been active in lodgepole pine ecoystems almost as long as lodgepole pine has existed. Frequency of infestations in a given area of forest appears to range from about 20 to 40 years, depending upon how rapidly some trees in the stand grow to large diameter and produce thick phloem, conditions conducive to buildup of beetle populations. In addition, trees must be at a latitude and elevation where temperatures are favorable for beetle development.

The Mountain Pine Beetle

The adult is stout, black to dark brown, cylindrical and about 6 mm long. The beetle usually completes one generation per year in lodgepole pine. However, two years may be required at high elevations and the cooler climates of northern latitudes. New adults emerge from the bark between late June and early September depending upon elevation, latitude, longitude, and weather conditions during the flight period. After a period of sparse, sporadic emergence, the majority of beetles emerge and make attacks in about a one-week period (Rasmussen, 1974). This rapid emergence by most of the population allows successful infestation of vigorous trees. If the attacking beetles are few in number, egg galleries may become impregnated with resin and all eggs and larvae are killed by resinosus (Reid et al., 1967). The tree may survive these light attacks.

The female initiates the attack, usually on the basal 2 m of the tree trunk, and produces an aggregating pheromone, *Trans* -verbenol (Pitman et al., 1968). This pheromone in conjunction with terpenes from the tree guides other beetles to the tree and serves as a signal for invasion of the host (Vité and Pitman, 1968). Beetles attack the tree en masse and kill it if their numbers are sufficient. To prevent overcrowding, attack density on individual trees is regulated by host condition (Renwick and Vité, 1970) and antiaggregative-rivalry pheromones that mask the aggregative pheromone (Rudinsky et al., 1974). The female usually mates early in gallery construction and lays eggs in irregularly alternating groups on the two sides of the vertical gallery. She lays about two eggs/cm of gallery; however, the number varies with size of female (Reid, 1962; McGhehey, 1971; Amman 1972a), with phloem thickness and temperature (Amman 1972a), and with freshness and moisture content of the bark (Reid, 1962). Eggs hatch in about two weeks and larvae feed individually in the inner bark (phloem). Larval galleries usually extend at right angles to the egg galleries, thereby girdling the tree.

Mature larvae excavate oval cells in the bark, lightly scouring the sapwood, where they pupate and later become adults. New adults feed within the bark prior to chewing exit holes through the outer bark and then emerge to attack healthy trees.

More females than males almost always survive. However, no single factor appears to be responsible for differential survival of the sexes. Differences have been attributed to crowding (W. E. Cole, 1973), length of cold storage (Watson, 1971; Safranyik, 1976), and phloem quality (Amman and Pace, 1976) in laboratory studies; and to drying in field studies (Amman and Rasmussen, 1974; Cole et al., 1976). The sexes survive about equally in large diameter trees where conditions appear most favorable to the beetle (Cole et al., 1976).

In addition to the girdling action of larvae, blue-stain fungi—*Cerato-stomella montia* (Rumbold, 1941) and *Europhium clavigerum* (Robinson-Jeffrey and Davidson, 1968)—are introduced by adult beetles and have been considered the primary cause of tree death (Safranyik et al., 1974). Fungal spores which probably are picked up during maturation feeding by the new adult are carried in a maxillary mycangium (Whitney and Farris, 1970), indicating a true symbiotic relationship of fungus and beetle. The spores are introduced into the tree as the beetles con-struct egg galleries. The blue-stain fungi invade the phloem, and es-pecially the sapwood of the xylem, where they interfere with conduction (Nelson, 1934). The principal benefit to the beetle appears to be regulation of moisture conditions in the tree during development. Trees having well developed blue stain dry out more rapidly than trees con-taining poorly developed blue stain following infestation, but remain more moist about 11 months following infestation when the beetle is completing development. Blue-stain fungi do not appear to be necessary to mountain pine beetle nutrition (Whitney, 1971).

E. clavigerum has been artificially inoculated into lodgepole pine to determine resistance to the fungus, and thus, an indicator of resis-tance to infestation by the beetle (Reid et al., 1967). The beetle killed more nonresistant than resistant trees rated according to response to fungal inoculation (Shrimpton and Reid, 1973). Trees rated potentially resistant had faster growth rates and thicker phloem than those rated nonresistant (Shrimpton, 1973).

Lodgepole Pine

Pinus contorta is one of the most widely distributed tree species in western North America, extending from the Yukon Territory to Baja California, and east to the Black Hills of South Dakota (Little, 1971). The lodgepole pine of concern here, *P. contorta* var. *latifolia*, is the inland variety found in mountainous areas from Colorado to the Yukon Territory.

Lodgepole pine grows rapidly where competition is limited, reaching a size of 24 m in height and 41 cm d.b.h. (diameter at breast height = 1.4 m above ground) in 50-60 years. Trees in unmanaged, even-aged stands on medium sites in Montana averaged 19 m in height and 21 cm d.b.h. at age 80 (Tackle, 1959). However, lodgepole pine on such sites is not mature until age 120, nor overmature until age 140 (Tackle, 1955).

Ecologically, lodgepole is typically described as seral, with low shade tolerance; possessing the ability to grow on almost any forest site;

having serotinous cones that require high temperatures to open and release seed; regenerating rapidly in large numbers that create stagnated stands; having rapid growth in young trees and slow growth in old trees; having high susceptibility to mistletoe infection and premature mortality from mountain pine beetle attack. Many of these characteristics contribute to large fuel buildups that lead to intense fires over large areas, thus renewing the cycle (Pfister and Daubenmire, 1975).

Pfister and Daubenmire (1975) recognized four basic successional roles for lodgepole pine:

1. Minor seral. Lodgepole pine is a minor component of young, even-aged, mixed species stands. It is rapidly replaced by shade-tolerant associates in 50-200 years; the more mesic the site, the sooner lodgepole pine is replaced.

2. Dominant seral. Lodgepole pine is the dominant cover type of even-aged stands with a vigorous understory of shade-tolerant species that will replace the lodgepole in 100-200 years. Succession occurs most rapidly where lodgepole pine and shade-tolerant associates become established simultaneously. Lodgepole pine gains dominance through rapid early growth, but shade-tolerant species persist and assume dominance as individual lodgepole pines die.

3. Persistent. Lodgepole pine forms the dominant cover type of even-aged stands with little evidence of replacement by shade-tolerant species. These species are present only as scattered individuals but apparently are too few and lack sufficient vigor to replace lodgepole pine. Lodgepole pine maintains its dominance because of inadequate seed sources for potential competitors, stand densities too great to allow regeneration of any other species, and light surface fires that remove seedlings without killing overstory lodgepole pine.

4. Climax. Lodgepole pine is the only species capable of growing on particular sites and is self-perpetuating. Some examples: In central Oregon, lodgepole pine forms an edaphic climax on poorly drained soils and a topoedaphic climax in frost pockets (Franklin and Dyrness, 1973). In Wyoming, lodgepole forms an edaphic climax on granitic soils in portions of the Bighorn Mountains (Despain, 1973) and on shallow, infertile soils of schist origin in portions of the Wind River mountains (Reed, 1976).

Mountain Pine Beetle-Lodgepole Pine Interactions

Many factors affecting beetle populations have been studied through life table sampling of populations using a method of bark removal outlined by Carlson and Cole (1965) and through systematic sampling of lodgepole pine stands described by Cole and Amman (1969). The four most important factors influencing beetle populations are structure of lodgepole pine stands, phloem thickness, moisture content of the tree during beetle development, and climate.

Infestations in Relation to Stand Structure

The mountain pine beetle infests and kills proportionately more large-than small-diameter trees. Hopping and Beall (1948) showed a 2% increase in mortality per cm increase in d.b.h. for stands in Alberta;

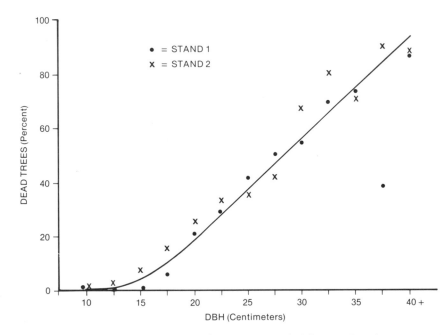

Fig. 1. Percent of lodgepole pine trees within each diameter class killed by the mountain pine beetle (Cole and Amman, 1969)

and Roe and Amman (1970) observed an increase of 3.5% in Wyoming and Idaho. Some of the greatest losses of lodgepole pine to the beetle occurred in the Big Hole Basin of Montana where 84% of the trees 23 cm and larger d.b.h. were killed (Evenden and Gibson, 1940). In two stands in northwestern Wyoming, mortality ranged from about 1% of trees 10 cm d.b.h. to 87% of those 41 cm and larger d.b.h. (Fig. 1) (Cole and Amman, 1969). Furthermore, the beetle attacks the trees of largest diameter each year of the infestation, until mostly small trees remain and the infestation then declines (Fig. 2) (Cole et al., 1976).

Shepherd (1966) studied behavior of the beetle in the laboratory and found that large dark objects against a light background were more attractive to beetles than small objects. His study indicates that the beetle uses visual stimuli, and selects trees to be attacked on the basis of size. Presently, this appears to be the most plausible explanation of the beetles' behavior. The evolutionary basis for this behavior is probably related to the much higher probability of encountering thick phloem (Fig. 3), the food supply of developing larvae (Amman, 1975).

Beetle Production in Relation to Phloem Thickness

Large diameter lodgepole pines, on the average, produce more mountain pine beetles per unit area of surface than do those of small diameter (Reid, 1963; Cole et al., 1976). The principal reason is the thicker phloem. Phloem thickness increases exponentially as diameter increases from 10 to 40 cm (Fig. 4). Furthermore, phloem thickness has been shown to be directly related to characteristics of good lodgepole pine vigor (D. M. Cole, 1973).

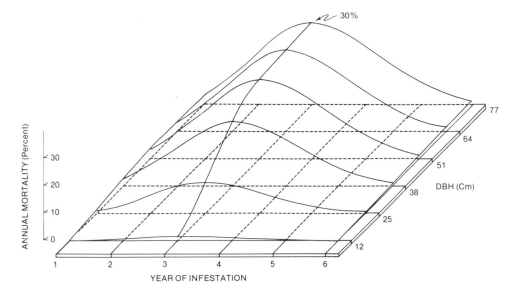

Fig. 2. Percent of lodgepole pine trees killed within each diameter class during each of the main years of a mountain pine beetle infestation (Cole et al., 1976)

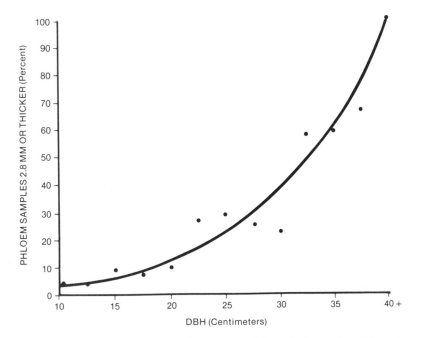

Fig. 3. Percent of phloem samples exceeding 2.8 mm by diameter class of lodgepole pine (Amman, 1975)

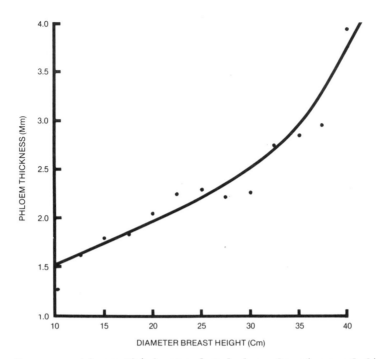

Fig. 4. Average phloem thickness for lodgepole pines of different
diameters

Laboratory rearings of beetles in lodgepole pine billets show average
production ranged from $3.0/dm^2$ for phloem 2 mm thick to $8.7/dm^2$ from
phloem 6 mm thick (Amman, 1972b). Greater beetle production has been
obtained with increased gallery densities. Production curves asymptote
at 22.7 cm of gallery/dm^2 in thin phloem and 26.0 cm/dm^2 in thick
phloem. These curves were maintained at higher gallery densities,
indicating that above these levels a constant production of beetles
could be expected in the laboratory (Amman and Pace, 1976). Laboratory
studies so far have failed to demonstrate a clear qualitative difference between phloem of young and old trees and between phloem of small
and large diameter trees. However, differences in sizes of beetles
reared from thick and thin phloem suggest a qualitative difference
(Amman and Pace, 1976).

Beetle Survival in Relation to Moisture Content of the Tree

Adequate moisture is essential throughout development of the mountain
pine beetle. Drying usually is greater in small diameter than in large
diameter trees infested by the beetle (Fig. 5), particularly in those
trees that had a slow rate of growth. Differential drying probably
accounts for some of the reduced beetle emergence (survival) observed
between large and small trees having similar phloem thickness (Cole,
1974, 1975).

The role of blue-stain fungi in regulating moisture content of the tree
is not completely clear. Reid (1961) observed that trees with abundant
blue-stain fungi were drier in the fall after attack than were trees
with poorly developed blue-stain fungi. I also observed this, but in

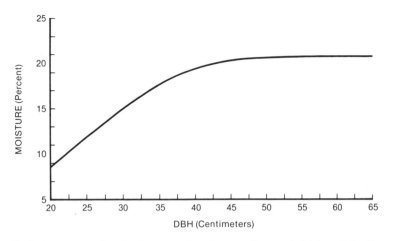

Fig. 5. Moisture content (percent of sample weight) of lodgepole pines 11 months after being infested and killed by the mountain pine beetle

addition found the opposite relation in early July, about 11 months following infestation. Trees having well-developed blue stain were more moist than trees in which blue stain was scarce. Beetle survival was low in trees with poorly developed blue stain. Blue-stain fungi appear to play a dual role—their presence results in increased drying in the fall and delayed drying in the spring. These conditions would be beneficial to both the fungus and the beetle.

Attack and gallery densities also influence rate of drying of the tree. Both increase over the several years of an infestation (Cole et al., 1976). Increased egg gallery density and increased numbers of feeding larvae in the phloem layer result in more rapid drying of the tree. The increase in attack and gallery densities has been attributed to a shift in the sex ratio of the beetle population toward more females (Cole et al., 1976). After most of the large diameter trees are killed and the beetles infest small diameter trees, drying is extensive and male survival declines. When these broods emerge and infest trees, probably not enough males are present to rapidly mate females and cause masking of the aggregative pheromone to limit the attacking population.

Beetle Infestations in Relation to Climate

Climate is a major limiting factor in the dynamics of the mountain pine beetle at extreme northern latitudes and at high elevations. Brood production by the beetle in bark of a given thickness is inversely related to elevation (Amman, 1969). With increased elevation, beetle development becomes so retarded that much of the beetle population enters the winter in stages particularly susceptible to being killed by cold temperatures—eggs and small larvae during the first winter, and prepupal larvae, pupae, and teneral adults during the second winter of the two-year life cycle at high elevations (Amman, 1973). Because of reduced brood survival, infestations are not as intense and fewer trees are killed as elevation increases (Fig. 6) (Amman and Baker, 1972; Amman et al., 1973).

Safranyik et al. (1974) outlined zones of infestation intensity for the mountain pine beetle in Canada, with the greatest intensity occurring at low elevations near the United States-Canada border. These zones

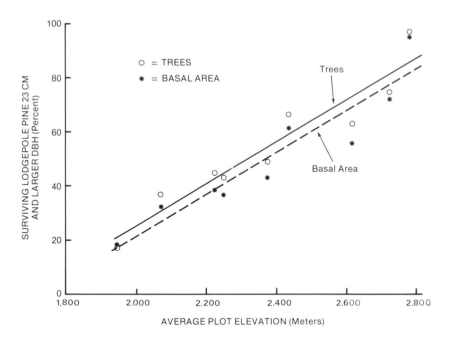

AVERAGE PLOT ELEVATION (Meters)

Fig. 6. Percent of lodgepole pine surviving at different elevations
after a mountain pine beetle infestation had subsided (Amman, 1975)

represent changes in infestation intensity at different latitude-ele-
vation combinations. Safranyik et al. (1975) have employed climatic
factors in their model to predict probability of beetle infestation.

Beetle Infestations in Relation to Habitat Type

The intensity of beetle infestations and subsequent numbers of lodgepole pine trees killed differ by habitat type (Roe and Amman, 1970).
A habitat type includes all sites with the potential of supporting the
same climax plant association (Daubenmire and Daubenmire, 1968).

Beetle activity was compared on three habitat types within the lodgepole pine type on the Teton and Targhee National Forests in northwestern Wyoming and southeastern Idaho: (1) Subalpine fir/dwarf vac-
cinium (*Abies lasiocarpa/Vaccinium scoparium*), or A/V type, generally found
at high elevations (range 1,997-2,576 m); (2) subalpine fir/mountain
lover (*A. lasiocarpa/Pachistima myrsinites*), or A/P type, generally found at
mid-elevations (range 2,042-2,377 m); and (3) Douglas-fir/pine grass
(*Pseudotsuga menziesii/Calamagrostis rubescens*), or P/C type, generally found
at low elevations (range 1,829-2,362 m) (Roe and Amman, 1970). Overlap
in these elevational ranges indicates that habitat typing is a better
way to classify stands than is a strictly elevational classification
for investigating the ecology of the beetle. A classification based on
habitat type considers environmental differences associated with slope,
aspect, soil, latitude, and other factors.

The A/P type had the largest proportion of infested stands and suffered the greatest amount of loss (Fig. 7). The second highest infestation
rate and tree losses were in stands on the P/C type, with the least

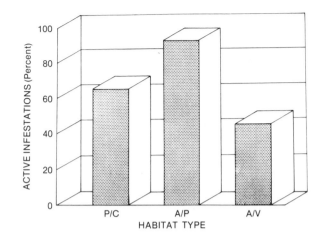

Fig. 7. Proportions of lodgepole pine stands in three habitat types that showed evidence of one or more mountain pine beetle infestations. *P/C:Pseudotsuga menziesii/Calamagrostis rubescens; A/P:Abies lasiocarpa/Pachistima myrsinites; A/V:Abies lasiocarpa/Vaccinium scoparium* (Roe and Amman, 1970)

infestation in the A/V type. Although temperature is important in differences among these habitat types, primarily because of differences in altitude, other factors enter in. For example, stands on more mesic sites will have trees that grow rapidly, and reach sizes and phloem thicknesses conducive to beetle population buildup more quickly than trees growing on more xeric sites. Consequently, beetle infestations will occur more frequently on sites providing for the best growth of lodgepole pine.

Role of the Mountain Pine Beetle in Lodgepole Pine Ecosystems

The role of the beetle differs in conjunction with the two basic ecological roles of lodgepole pine—where lodgepole pine is seral and where it is persistent or climax. The beetles' continued role in the seral stands will depend upon the presence of fire.

Role of Mountain Pine Beetle Where Lodgepole Pine Is Seral

Absence of Fire

Lodgepole pine stands depleted by the beetle and not subjected to fire are eventually succeeded by the more shade-tolerant species consisting primarily of Douglas-fir at the lower elevations and subalpine fir and Englemann spruce at the higher elevations throughout most of the Rocky Mountains (Fig. 8). Starting with a stand generated by fire, lodgepole pine grows at a rapid rate and occupies the dominant position in the stand. Fir and spruce seedlings also established in the stand grow more slowly than lodgepole pine.

With each infestation, the beetle kills most of the large, dominant lodgepole pines. After the infestation, both residual lodgepole pine

Fig. 8. Subalpine fir and Douglas-fir seedlings growing in openings
created when mountain pine beetles killed some of the larger dominant
lodgepole pines (trees on the ground)

and the shade-tolerant species increase their growth. When the lodgepole pines are of adequate size and phloem thickness, another beetle
infestation occurs. This cycle is repeated at 20-40 year intervals
depending upon growth of the trees, until lodgepole pine is eliminated
from the stand.

Increment cores taken from subalpine fir trees growing within lodgepole
pine stands in northwestern Wyoming show growth release at approximately 20-year intervals (Fig. 9). The more recent releases were correlated with periods of beetle activity, but there was no way of re-
lating the older release periods to infestations. However, weather
records from nearby stations indicated that the earlier release periods were not related to abundant moisture; in fact, several occurred
when moisture was deficient. Consequently, increased growth was the
result of stand disturbance, the most likely being an infestation of
mountain pine beetle (Roe and Amman, 1970). Evidence of older beetle
infestations consisting of the typical egg gallery etchings in the
sapwood was found on fallen trees. These fallen trees could not be
dated because of decay.

Subalpine fir succession in three lodgepole pine stands is shown in
Figure 10 (Roe and Amman, 1970). The Moody Meadows stand is in the
early stages of succession. The subalpine fir understory consists of
only a few trees per hectare, most of which have small diameters.
However, in this stand 2754 subalpine seedlings less than 2.5 cm d.b.h.
per hectare were present. These will grow to fill overstory openings
created by future beetle infestations.

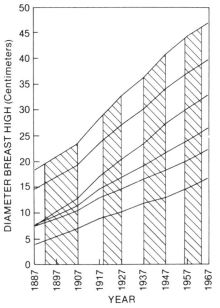

Fig. 9. Trends in diameter growth of
subalpine fir trees in a stand of
lodgepole pine that had been subjected to four mountain pine beetle
infestations. Crosshatched bars were
periods of beetle infestation
(Roe and Amman, 1970)

The Pilgrim Mountain stand represents a more advanced stage of succession. Both subalpine fir, which was on the cool mesic sites, and Douglas-
fir, which was on the warmer xeric sites, were present in this stand. A
large reservoir of 6945 fir seedlings less than 2.5 cm d.b.h. per hec-
tarc was present, ready to assume a more prominent position in the
stand as lodgepole pine trees are killed by the beetle.

In the Dell Creek stand, succession by subalpine fir is almost com-
pleted. In spite of the small number of large lodgepole pines remaining
in this stand, the beetle was able to locate and infest them. Trends
typical of succeeding species are very apparent in these data, with
large numbers of small fir trees declining to a few large trees, some
of which have reached 41 cm or larger d.b.h. Data from lodgepole pine
stands located at lower elevations indicate a similar relationship
with Douglas-fir.

The role played by the mountain pine beetle in stands where lodgepole
pine is seral is to periodically remove the large, dominant pines. This
provides growing space for subalpine fir and Douglas-fir, thus has-
tening succession by these species. The continued presence of the beetle in these mixed-species stands is as dependent upon fire as that of
lodgepole pine, without it both are eliminated.

Presence of Fire

Where lodgepole pine is seral, forests are perpetuated through the
effects of periodic fires (Tackle, 1964). Fires tend to eliminate competitive tree species such as Douglas-fir, the true firs, and spruces.
Following fire, lodgepole pine usually seeds in abundantly. Serotinous
cones attached to the limbs of the tree open because of the intense
heat of the fire and release their seed (Clements, 1910; Lotan, 1975).

Large accumulations of dead material caused by periodic beetle infestations result in very hot fires when they do occur (Brown, 1975). Hot
fires of this nature eliminate Douglas-fir, which otherwise is more

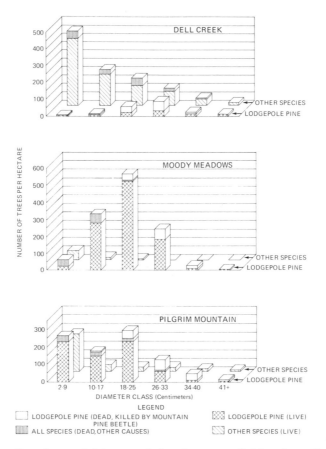

Fig. 10. Distribution of live and dead trees following the most recent infestation by the mountain pine beetle in three lodgepole pine stands in northwest Wyoming (Dell Creek and Pilgrim Mountain) and southeast Idaho (Moody Meadows) (Roe and Amman, 1970)

resistant to fire damage than lodgepole pine. The dominant shade-tolerant species are eliminated, resulting in a return to a pure lodgepole pine forest. On the other hand, light surface fires would not be adequate to kill large, thick-barked Douglas-fir and return lodgepole pine to a dominant position in the stand.

Following regeneration of lodgepole pine after fire, the mountain pine beetle-lodgepole interactions would be similar to those described in the absence of fire. A fire may interrupt the sere at any time, reverting the stand back to pure lodgepole pine. However, once succession is complete lodgepole pine seed will no longer be available to seed the burned areas except along edges where the spruce-fir climax joins persistent or climax lodgepole pine.

Role of Mountain Pine Beetle Where Lodgepole Pine Is Persistent or Climax

Lodgepole pine is persistent over large acreages and because of the number of shade-tolerant individuals of other species found in such persistent stands, the successional status is unclear (Pfister and Daubenmire, 1975). In any case, lodgepole pine persists long enough for a number of beetle infestations to occur. In such cases and those of a more limited nature when lodgepole pine is climax because of special climatic or soil conditions, the forest consists of trees of different sizes and ages ranging from seedlings to a few overmature individuals. In these forests, the beetle infests and kills most of the lodgepole pines as they reach larger sizes. Openings created in the stand as a result of the larger trees being killed, are seeded by lodgepole pine. The cycle is then repeated as other lodgepole pines reach sizes and phloem thicknesses conducive to increases in beetle populations (Fig. 11).

The result is two- or three-story stands consisting of trees of different ages and sizes. A mosaic of small clumps of different ages and sizes may occur. The overall effect is likely to be more chronic infestations by the beetle because of the more constant source of food. Beetle infestations in such forests may result in death of fewer trees per hectare during each infestation than would occur in even-aged stands developed after fires and in those where lodgepole pine is seral.

Fig. 11. Openings created when the mountain pine beetle kills large dominant trees in persistent and climax lodgepole pine stands are seeded by lodgepole pine. Stump is remnant of tree killed by mountain pine beetle about 12 years ago

Fires in persistent and climax lodgepole pine forests should not be
as hot as those where large epidemics of beetles have occurred. Smaller,
more continuous deposits of fuel are available on the forest floor. The
lighter beetle infestations, and thus lighter accumulations of fuel,
would result in fires that would eliminate some of the trees but probably would not cause total regeneration of the stand. This would be
beneficial to the beetle because a more continuous supply of food would
be maintained. Where large accumulations of fuel occur after large beetle epidemics, fire would completely eliminate the beetles' food supply
from vast acreages for many years while the entire stand of trees grew
from seedlings to sizes conducive to beetle infestation.

The mountain pine beetle's evolutionary strategies have been successful.
It has exploited a niche that no other bark beetle has been able to
exploit, that of harvesting lodgepole pine trees as they reach or
slightly before they reach maturity. Such trees are at their peak as
food for the beetle. Harvesting at this time in the age of the stand
maintains the vigor of the stand, and keeps the stand at maximum productivity.

References

Amman, G.D.: Mountain pine beetle emergence in relation to depth of
    lodgepole pine bark. USDA For. Serv. Res. Note INT-96, 1969, 8 p.
Amman, G.D.: Some factors affecting oviposition behavior of the moun-
    tain pine beetle. Environ. Entomol. 1,691-695 (1972a)
Amman, G.D.: Mountain pine beetle brood production in relation to thick-
    ness of lodgepole pine phloem. J. Econ. Entomol. 65,138-140 (1972b)
Amman, G.D.: Population changes of the mountain pine beetle in relation
    to elevation. Environ. Entomol. 2,541-547 (1973)
Amman, G.D.: Insects affecting lodgepole pine productivity. In: Man-
    agement of Lodgepole Pine Ecosystems, Symp. Proc. Baumgartner, D.M.
    (ed.). Pullman, Wash: Wash. State Univ., 1975, pp. 310-341
Amman, G.D., Baker, B.H.: Mountain pine beetle influence on lodgepole
    pine stand structure. J. Forestry 70,204-209 (1972)
Amman, G.D., Baker, B.H., Stripe, L.E.: Lodgepole pine losses to moun-
    tain pine beetle related to elevation. USDA For. Serv. Res. Note
    INT-171, 1973, 8 p.
Amman, G.D., Pace, V.E.: Optimum egg gallery densities for the mountain
    pine beetle in relation to lodgepole pine phloem thickness. USDA For.
    Serv. Res. Note INT-209, 1976, 8 p.
Amman, G.D., Rasmussen, L.A.: A comparison of radiographic and bark-
    removal methods for sampling of mountain pine beetle populations.
    USDA For. Serv. Res. Pap. INT-151, 1974, 11 p.
Brown, J.K.: Fire cycles and community dynamics in lodgepole pine
    forests. In: Management of Lodgepole Pine Ecosystems, Symp. Proc.
    Baumgartner, D.M. (ed.) Pullman, Wash., Wash. State Univ., 1975,
    pp. 429-456
Carlson, R.W., Cole, W.E.: A technique for sampling populations of the
    mountain pine beetle. USDA For. Serv. Res. Pap. INT-20, 1965, 13 p.
Clements, F.E.: The life history of lodgepole pine burn forests. U.S.
    Dep. Agric., For. Serv. Bull. 79, 1910, 56 p.
Cole, D.M.: Estimation of phloem thickness in lodgepole pine. USDA
    For. Serv. Res. Pap. INT-148, 1973, 10 p.
Cole, W.E.: Crowding effects among single-age larvae of the mountain
    pine beetle, *Dendroctonus ponderosae* (Coleoptera: Scolytidae). Environ.
    Entomol. 2,285-293 (1973)
Cole, W.E.: Competing risks analysis in mountain pine beetle dynamics.
    Res. Popul. Ecol. 15,183-192 (1974)

Cole, W.E.: Interpreting some mortality factor interactions within
    mountain pine beetle broods. Environ. Entomol. 4,97-102 (1975)
Cole, W.E., Amman, G.D.: Mountain pine beetle infestations in relation
    to lodgepole pine diameters. USDA For. Serv. Res. Note INT-95, 1969,
    7 p.
Cole, W.E., Amman, G.D., Jensen, C.E.: Mathematical models for the
    mountain pine beetle-lodgepole pine interaction. Environ. Entomol.
    5,11-19 (1976)
Daubenmire, R., Daubenmire, J.B.: Forest vegetation of eastern Wash-
    ington and northern Idaho. Wash. Agr. Exp. Stn. Tech. Bull. 60,
    104 p (1968)
Despain, D.G.: Vegetation of the Bighorn Mountains of Wyoming in re-
    lation to substrate and climate. Ecol. Monogr. 43,329-355 (1973)
Evenden, J.C., Gibson, A.L.: A destructive infestation in lodgepole
    pine stands by the mountain pine beetle. J. Forestry 38,271-275
    (1940)
Franklin, J.F., Dyrness, C.T.: Natural vegetation of Oregon and Wash-
    ington. USDA For. Serv. Gen. Tech. Rep. PNW-8, 1973, 417 p.
Hopping, G.R., Beall, G.: The relation of diameter of lodgepole pine to
    incident of attack by the bark beetle *Dendroctonus monticolae* Hopk.
    Forestry Chron. 24,141-145 (1948)
Little, E.L., Jr.: Atlas of United States trees. I. Conifers and Im-
    portant Hardwoods. U.S. Dep. Agr. Misc. Publ. 1146, Vol. I, 9 p.,
    200 maps (1971)
Lotan, J.E.: The role of cone serotiny in lodgepole pine forests. In:
    Management of Lodgepole Pine Ecosystems, Symp. Proc. Baumgartner,
    D.M. (ed.) Pullman, Wash: Wash. State Univ., 1975, pp. 471-495
McGhehey, J.H.: Female size and egg production of the mountain pine
    beetle, *Dendroctonus ponderosae* Hopkins. Can. For. Serv., North. For.
    Res. Cent., Inform. Rep. NOR-X-9, 1971, 18 p.
Nelson, R.M.: Effect of bluestain fungi on southern pines attacked by
    bark beetles. Phytopathology 7,327-353 (1934)
Pfister, R.D., Daubenmire, R.: Ecology of lodgepole pine, *Pinus contorta*
    Dougl. In: Management of Lodgepole pine Ecosystems, Symp. Proc.
    Baumgartner, D.M. (ed.). Pullman, Wash: Wash. State Univ., 1975,
    pp. 27-46
Pitman, G.B., Vité, J.P., Kinzer, G.W., Fentiman, A.F., Jr.: Bark
    beetle attractants: *Trans*-verbenol isolated from *Dendroctonus*. Nature
    (London) 218,168-169 (1968)
Rasmussen, L.A.: Flight and attack behavior of mountain pine beetles
    in lodgepole pine of northern Utah and southern Idaho. USDA For.
    Serv. Res. Note INT- 180, 1974, 7 p.
Reed, R.M.: Coniferous forest habitat types of the Wind River Mountains,
    Wyoming. Am. Midl. Naturalist 95,159-173 (1976)
Reid, R.W.: Moisture changes in lodgepole pine before and after attack
    by the mountain pine beetle. Forestry Chron. 37,368-403 (1961)
Reid, R.W.: Biology of the mountain pine beetle, *Dendroctonus monticolae*
    Hopkins, in the east Kootenay region of British Columbia. II. Be-
    havior in the host, fecundity, and internal changes in the female.
    Can. Entomologist 94,605-613 (1962)
Reid, R.W.: Biology of the mountain pine beetle, *Dendroctonus monticolae*
    Hopkins, in the east Kootenay region of British Columbia. III. In-
    teraction between the beetle and its host, with emphasis on brood
    mortality and survival. Can. Entomologist 95,225-238 (1963)
Reid, R.W., Whitney, H.S., Watson, J.A.: Reactions of lodgepole pine
    to attack by *Dendroctonus ponderosae* Hopkins and blue stain fungi. Can.
    J. Botany 45,1115-1126 (1967)
Renwick, J.A.A., Vité, J.P.: Systems of chemical communication in
    Dendroctonus. Contrib. Boyce Thompson Inst. Plant Res. 24,283-292 (1970)
Robinson-Jeffrey, R.C., Davidson, R.W.: Three new *Europhium* species with
    Verticicladiella imperfect states on blue-stained pine. Can. J. Botany
    46,1523-1527 (1968)

Roe, A.L., Amman, G.D.: The mountain pine beetle in lodgepole pine
    forests. USDA For. Serv. Res. Pap. INT-71, 1970, 23 p.
Rudinsky, J.A., Morgan, M.E., Libbey, L.M., Putnam, T.B.: Antiaggrega-
    tive rivalry pheromone of the mountain pine beetle, and a new arrest-
    ant of the southern pine beetle. Environ. Entomol. 3,90-98 (1974)
Rumbold, C.T.: A blue stain fungus, *Ceratostomella montium* N. sp., and
    some yeasts associated with two species of *Dendroctonus*. J. Agr. Res.
    62,589-601 (1941)
Safranyik, L.: Size- and sex-related emergence, and survival in cold
    storage, of mountain pine beetle adults. Can. Entomologist 108,209-212
    (1976)
Safranyik, L., Shrimpton, D.M., Whitney, H.S.: Management of lodgepole
    pine to reduce losses from the mountain pine beetle. Can. Dep.
    Environ., For. Serv., Pac. For. Res. Cent. Tech. Rep. 1, 1974, 24 p.
Safranyik, L., Shrimpton, D.M., Whitney, H.S.: An interpretation of the
    interaction between lodgepole pine, the mountain pine beetle and its
    associated blue stain fungi in western Canada. In: Management of
    Lodgepole Pine Ecosystems, Symp. Proc. Baumgartner, D.M. (ed.).
    Pullman, Wash: Wash. State Univ., 1975, pp. 406-428
Shepherd, R.F.: Factors influencing the orientation and rates of
    activity of *Dendroctonus ponderosae* Hopkins (Coleoptera: Scolytidae).
    Can. Entomologist 98,507-518 (1966)
Shrimpton, D.M.: Age- and size-related response of lodgepole pine to
    inoculation with *Europhium clavigerum*. Can. J. Botany 51,1155-1160 (1973)
Shrimpton, D.M., Reid, R.W.: Change in resistance of lodgepole pine
    to mountain pine beetle between 1965 and 1972. Can. J. For. Res.
    3,430-432 (1973)
Tackle, D.: A preliminary stand classification for lodgepole pine in
    the Intermountain Region. J. Forestry 53,566-569 (1955)
Tackle, D.: Silvics of lodgepole pine. U.S. Dep. Agric., For. Serv.,
    Intermt. For. and Range Exp. Stn., Misc. Publ. 19, 1959, 24 p.
Tackle, D.: Ecology and silviculture of lodgepole pine. Soc. Am. For.
    Proc. 1964,112-115
Vité, J.P., Pitman, G.B.: Bark beetle aggregation: effects of feeding on
    the release of pheromones in *Dendroctonus* and *Ips*. Nature (London)
    218,169-170 (1968)
Watson, J.A.: Survival and fecundity of *Dendroctonus ponderosae* (Coleoptera:
    Scolytidae) after laboratory storage. Can. Entomologist 103,1381-1385
    (1971)
Whitney, H.S.: Association of *Dendroctonus ponderosae* (Coleoptera: Scoly-
    tidae) with blue stain fungi and yeasts during brood development
    in lodgepole pine. Can. Entomologist 103,1495-1503 (1971)
Whitney, H.S., Farris, S.H.: Maxillary mycangium in the mountain pine
    beetle. Science 167,54-55 (1970)

Chapter 2

The Significance of Phytophagous Insects in the *Eucalyptus* Forests of Australia

P. A. MORROW

Introduction

Ninety-two percent of the forests and woodlands of Australia are dominated by evergreen hardwoods of the genus *Eucalyptus* (Myrtaceae), an
endemic genus of more than 600 species (Pryor and Johnson, 1971). Apart
from this dominance of the forests by a single genus, there are at
least three related aspects of Australian forests which are of considerable ecological interest: (1) *Eucalyptus* species frequently have
sharply defined, narrow geographical ranges which are closely associated with local environmental factors such as aspect (Pook and
Moore, 1966), drainage (Florence, 1971), and especially, changes in
soil chemistry (Beadle, 1954; C. Moore, 1959; Winterhalter, 1963;
Parsons and Specht, 1966). (2) Despite their apparent specialization
to particular edaphic or microclimatic conditions, *Eucalyptus* species
rarely form monospecific stands. Instead, species tend to have over-
lapping ranges so that any particular set of environmental conditions
is usually characterized by a stable mixture of from two to five
species (Pryor, 1959). Eucalypt forests are, therefore, complex mosaics
of quite sharply delineated plant associations. Pryor (1953) showed
that as many as 12 different associations may be present in one square
mile of forest. The apparently stable coexistence of closely related
species that seem to exploit the same resources appears to contradict
the principle of competitive exclusion (Hardin, 1960; Hutchinson,
1966). (3) Phytophagous insects in *Eucalyptus* forests and woodlands
cause high, chronic levels of damage. Defoliation is estimated to be
20-50% of the leaf area (Burdon and Chilvers, 1974a; Kile, 1974; Fox,
pers. comm., Morrow, unpubl.). Moreover, outbreaks of herbivorous
insects are frequent and some may continue for longer than ten years
(Campbell, 1962; Clark and Dallwitz, 1975).

This paper contains a comprehensive review of the impact of insects
on seed production and establishment and the growth and mortality of
sapling and mature eucalypts. The relationships between host specificity of the herbivores, the productivity of coexisting eucalypts,
and the possible roles of insects in preventing competitive exclusion
in associations of closely related *Eucalyptus* species are explored.

Natural History of *Eucalyptus*

Taxonomically, *Eucalyptus* is a difficult genus which contains large groups
of morphologically similar species that often hybridize under natural
conditions. The genus has recently been revised and divided into five
subgenera (Pryor and Johnson, 1971). Subgenera consist of closely
related species which can hybridize with one another, but basic breeding incompatibilities exist between subgenera (Pryor, 1959). The nomenclature of *Eucalyptus* in this paper follows the classification of
Pryor and Johnson (1971).

Eucalypts are pollinated by insects and birds and although they usually outcross, most trees are self-fertile. *Eucalyptus* seeds are small (1-3 mm on average) and only a small proportion are viable. Larsen (1965) and Jacobs (1955) estimated that, for eucalypts in general, only 10-15% of ovules are fertilized. After release from the fruit capsule, fertile seed germinates immediately if conditions are suitable but loses its viability within 14 months on the ground (Cremer, 1965; Dexter, 1967) so that a "seed bank" cannot develop in the soil (Creamer and Mount, 1965; Howard and Ashton, 1967).

The vast majority of eucalypts are evergreen. Leaves may be retained for 18 months on average (range: six months to three years). Flowering, fruiting, insect attack, growth flushes and weather all affect leaf longevity (Jacobs, 1955). The system of bud production in *Eucalyptus* is particulary relevant in considering the effect of defoliating insects on the trees. Eucalypts have four ways of producing leafy shoots (Jacobs, 1955).

1. Rapid crown development results from the presence of a stalked, naked bud (i.e. without protective scales), which is present in the axil of every leaf. The naked bud is exposed as soon as the subtending leaf unfolds and develops a new shoot concurrently with the parent shoot. Each leaf on these first order shoots also has a naked bud which gives rise to a second order shoot and so forth up to four orders.

2. Meristematic tissue which is located in each leaf axil may give rise to "accessory" buds and shoots if the parent shoot and naked buds are removed. Accessory buds have the same growth potential as the naked buds and the meristematic area is capable of replacing buds which are destroyed indefinitely (Penfold and Willis, 1961).

3. The meristematic areas in leaf axils also give rise to "epicormic" buds. When the original leaf is shed, the meristematic area forms a strand which has the potential to produce buds: this strand grows radially outwards at the same rate as the cambium so that incipient bud tissue is present just beneath the most recently formed wood on the trunk and branches. The epicormic buds produce leafy shoots if the crown is severely damaged, permitting rapid recovery from crown fires and defoliation.

4. Most eucalypts have a fourth source of buds which develop from a heavily lignified swelling at the base of the trunk. This "lignotuber" grows throughout the life of a tree—often downward so that it enters the soil. The lignotuber contains dormant bud strands and, if the stem is killed, it produces numerous "coppice" shoots, one or more of which eventually become dominant and regenerate the aerial parts of the tree (Jacobs, 1955; Penfold and Willis, 1961). After severe fires, coppicing from lignotubers is the chief form of regeneration.

Effect of Insects on Seed Production and Establishment

Because there is no soil bank of seed, population recruitment is dependent on the store of viable seed held on the trees. Insects can affect the size of this store by destroying flower buds (Grose, 1957; Campbell, 1962; Dexter, 1967) and by attacking seed held in the fruit capsule. Insects commonly destroy 15-20% of the seed in *Eucalyptus delegatensis* (Grose, 1957) and about 5% of the viable seed in 68 Victorian eucalypt species (Grose and Zimmer, 1958).

Seed losses to insects are much higher after seed is shed. For example, ants and lygaeid buds removed 80-100% of seed occurring on the ground in less than two weeks (Grose, 1957; Cremer, 1966; Ashton, 1975). Treatment of experimental plots with insecticide caused seedling germination to increase by at least 50% for several species of *Eucalyptus* (Jacobs, 1955; Gilbert, 1959).

During the first weeks after seed germination, factors such as frost, drought, and fungal attack account for most seedling mortality (Gilbert, 1959; Dexter, 1967). However, once seedlings are established insect attack becomes an increasingly important factor (Dexter, 1967; Greaves, 1967).

Effects of Defoliation of Growth and Survival

Natural Defoliation

Several studies emphasize that the impact of grazing depends on the timing of defoliation, as well as on the amount of tissue removed. This is particularly well illustrated by studies on scarab beetles (Carne et al., 1974), chrysomelid beetles (Greaves, 1966) and phasmatids (Shepherd, 1957; Mazanec, 1967; Readshaw and Mazanec, 1969). Scarabs caused significant reductions in growth, but no mortality, even if they removed more than 60% of the leaf tissue. Nominal damage resulted because the scarabs were active in early spring when the effects of defoliation are minimal and they restricted their attacks to fully expanded leaves of the current season, ignoring buds, immature leaves, and older senescent leaves. However, if in addition to scarab attack, other insect species attacked buds and expanding tissues, severe growth reduction and some mortality resulted.

The response of *Eucalyptus regnans* to the chrysomelid beetle, *Chrysophtharta bimaculata*, also depends on time of defoliation. Defoliation late in the growing season inhibited growth more than defoliation early in the season. On the average, trees sprayed with insecticide had height growth which was double that of untreated trees (Greaves, 1966).

The phasmatid, *Didymuria violescens*, dramatically reduces the growth of *E. delegatensis* and *E. regnans* because defoliation continues for practically the entire growing season (Mazanec, 1966). Defoliation by phasmatids resulted in about 20% less diameter growth (Readshaw and Mazanec, 1969). These *Eucalyptus* species are able to survive repeated partial defoliation (Readshaw and Mazanec, 1969), but one complete defoliation may cause 59-72% mortality (Shepherd, 1957).

The jarrah leaf miner, *Perthida glyphopa*, retards the growth of mature *Eucalyptus marginata* and *E. rudis*, but does not kill them. For example, trees infested with leaf miners had diameter increments which were only 33 and 83% of that for noninfested trees. Growth reduction was proportional to the intensity of infestation (Mazanec, 1974). The growth reduction may be due not only to the removal of tissue by the leaf miner but also to the premature dropping of infested leaves. Leaves are normally retained for a least two years, but when heavily infested they may be shed after only five months. Lightly damaged leaves may fall before they are one year old. When trees are heavily infested, the crowns die back and new tissue is produced from epicormic buds.

In inland woodlands of southeastern Australia, adults and nymphs of the psyllid, *Cardiaspina albitextura*, feed on mature leaves of *Eucalyptus blakelyi*, inducing a condition resembling advanced physiological senescence. When heavily attacked, the tree sheds many damaged leaves (Clark, 1962). If individuals are defoliated at intervals of less than three years, the crown dies back and produces progressively less foliage. In some areas, outbreaks of psyllids persisted for more than 10 years and heavy tree mortality eventually resulted (Clark and Dallwitz, 1975).

Experimental Defoliation

The timing of defoliation is critical because the plant's ability to tolerate and recover from defoliation varies with its levels of energy reserves. Starch reserves are apparently highest just before the growing season commences and are rapidly depleted as starch is used for leaf production and growth. After leaf flushing, starches begin to accumulate again in the sapwood (Bamber and Humphreys, 1965). Bamber and Humphreys (1965) suggested that the effects of defoliation are more severe during periods of rapid growth when starch is being rapidly depleted, and plant death occurs when starch reserves are reduced to a level below that required for growth processes in the absence of photosynthetic tissue. This suggestion is supported by the data of Cremer (1973) who found that ability of several *Eucalyptus* species to recover from total defoliation was maximal in late winter or in spring before growth began and minimal at the end of the growing season. Furthermore, the response to defoliation was greatly influenced by growth conditions. Seedlings growing on dry sites were usually able to survive defoliation except when it occurred in late autumn. However, in wetter areas, where seedlings grew rapidly, defoliation resulted in high mortality at most times of year (Cremer, 1973). Stem starch content of rapidly growing control seedlings was very low or totally depleted from late spring until autumn and was highest in late winter-early spring, the only time when seedlings on wet sites were able to recover from defoliation. On dry sites, seedlings had high starch levels throughout most of the year (Cremer, 1973). Therefore, it appears that rapidly growing seedlings are unable to recover from defoliation because they lack adequate reserves to refoliate. Conversely, on dry sites, reserves are never totally depleted because environmental constraints limit the rate at which carbohydrate is used and, consequently, plants have adequate starch reserves available for refoliation.

The amount of defoliation also affects plant response. Cremer (1972) examined the effects of partial defoliation and bud removal on height growth of *E. regnans*. When all leaves or all buds were removed once from the upper 20% of seedling crowns, there was little or no growth reduction. However, if auxiliary buds were removed weekly, growth losses were much greater, and if the apical bud was also removed, height growth was severely retarded. Carne et al. (1974) artificially defoliated saplings of *Eucalyptus grandis* by removing only fully expanded leaves of the current season's growth, the tissue normally eaten by scarab beetles. Buds, currently expanding leaves, and leaves from the previous season were not removed. When 90-95% of the fully expanded, current leaves were removed, mean height growth was 61, 32 and 48% of controls when defoliation occurred in spring, early summer, and midsummer, respectively. Seventy-five to 80% leaf removal had less impact— the greatest growth depression (30%) occurred after spring defoliation and the least (9%) after midsummer defoliation. Three serial defoliations of 30% in spring, late spring, early summer had no growth impact. Mazanec (1966) studied the effects of both artificial and natural defoliation on diameter growth in *E. delegatensis*, a species noted for its sensitivity to defoliation. He removed both mature and very young leaves from the upper third of crowns, or completely defoliated sap-lings in early summer or in autumn. The treatments were repeated every year or in alternate years. Eighty-three percent of the saplings were killed by a complete defoliation at the end of the main growing period whereas all trees survived complete defoliations in the autumn. Growth rates of surviving trees declined with increasing severity of defolia-tion and no saplings survived two complete defoliations regardless of whether they were defoliated in successive or alternate years.

In summary, if defoliation occurs after spring growth has depleted starch reserves, the plant is less able to recover than if defolia-tion occurs later in the season when starch reserves have increased. Reduction in height and diameter growth increase with increasing levels of defoliation. If buds as well as leaves are removed (Cremer, 1972), and defoliation is continuous (Bamber and Humphreys, 1965; Greaves, 1966; Mazanec, 1966), growth is severely depressed, and mortality becomes important. Certain species, such as *E. regnans*, *E. delegantensis*, and *Eucalyptus obliqua*, are very sensitive to defoliation (Mazanec, 1966; Greaves, 1967). As a result, the levels of mortality for these species are atypical for the genus as a whole. Studies on other eucalypts indicate higher rates of survival for severely defoliated trees (Moore, 1959; Campbell, 1962; Carne, 1965; Wallace, 1970; Mazanec, 1974; Clark and Dallwitz, 1975).

Influence of Insects on the Composition of Plant Associations

Because phytophagous insects can have a major impact on the product-ivity of eucalypts, they have the potential to affect community struc-ture and species composition. However, for insects to exert such in-fluences, it is necessary that they feed differentially on *Eucalyptus* species within a single association, or that response of the plants to insect feeding are different. Differential impact on plant productivity may result when insects (1) restrict their attack to only one or a few of the available host species within a community, or (2) exhibit a density-dependent response such that the relative severity of their attacks on a particular plant species increase with the density of that species.

Host Preferences of Phytophagous Insects

All of the insects discussed so far show some degree of host prefer-ance, feeding on only a few of the many plant species available to them. For instance, several species of chrysomelid beetle have large geographic ranges and occur in different eucalypt associations where their preference depends on the particular species present. In the Snowy Mountains near Canberra, *Chrysophtharta bimaculata* appears to spec-ialize on the dominant eucalypt, *E. pauciflora* (Morrow, 1977) where-as in Tasmania it feeds upon *E. regnans*, *E. obliqua* and *E. delegatensis* (Greaves, 1966, Kile, 1974, de Little and Madden, 1975). *C. bimaculata* concentrates on one or two closely related species within an associ-ation, although it seems able to feed on other species (de Little and Madden, 1975). Other examples of site-dependent preferences are found among the phasmatids (Hadlington and Hoschke, 1959), sawflies (Carne, 1965), leaf miners (Campbell, 1962; Wallace, 1970; Mazanec, 1974), leaf-sucking psyllids (Moore, 1970), scarabs and other paropsine bee-tles (Carne, 1968; Carne et al., 1974; Fox, pers. comm.).

When insect numbers are high, however, less preferred species may be attacked. Carne (1965) found that the sawfly, *Perga affinis*, normally prefers *E. melliodora*, *E. blakelyi*, and *Eucalyptus camaldulensis*. But, when population densities are high and oviposition sites on all trees of these species are saturated, it will attack *Eucalyptus bridgesiana* and *Eucalyptus rubida*, if they are close in proximity to the preferred hosts. On the other hand, *Eucalyptus rossii* and *Eucalyptus woollsiana* were never attacked even though they frequently occurred with the most favored hosts. Similarly, the psyllid, *Cardiaspina albitextura*, feeds primarily on *E. blakelyi* but will attack a closely related species, *E. camaldulensis*, when its densities are very high. Commonly associated trees, *Eucalyptus melliodora*, *Eucalptus polyanthemos*, and *Eucalyptus albens* are never attacked by *C. albitextura*; although each of these eucalypts is attacked by different psyllid species (Clark, 1962).

Carne et al. (1974) studied insect defoliation in plantations of *E. grandis*, *Eucalyptus saligna*, and *Eucalyptus pilularis* in an area where all three species grow naturally. Two scarab beetles, *Anoplognathus porosus* and *A. chloropyrus*, restricted their attack entirely to *E. grandis*. However, *A. chloropyrus* ignored *E. grandis* when *E. dunni*, which occurs outside the insect's geographic range, was planted near *E. grandis*.

It is clear that preference for one or a few hosts within an association is common among *Eucalyptus* herbivores and that this specificity is not just a question of relative abundance of either the insects or the plants since many eucalypts are never attacked by particular insect species. Of the group of plants acceptable to an insect within a particular association, some species are preferred to others. Even when host preference is very strong, other acceptable plants may be attacked if infestations are very dense on the preferred host. This overflow is a function of both insect population density and the proximity of the preferred host to less acceptable hosts. Because the insects are absent from areas lacking the preferred host but containing many of the less acceptable eucalypts, it appears that utilization of the less acceptable host does not involve a change of preference. In extreme situations, an insect may change its hosts in response to changes in the host availability spectrum. This was the case for the scarab beetle which switched from its normal, preferred host to a eucalypt newly introduced to an area (Carne et al., 1974).

Effects of Host Preferences on Eucalypt Associations

When abundant insect species restrict their attacks to a subset of the eucalypt species within an association, they have the potential to affect the relative abundances of these eucalypts by impairing their abilities to compete with other eucalypts. The same effect may result from the combined attack of numerous species of insects on a particular eucalypt species. There is abundant anecdotal evidence which suggests that insect damage is consistently high on many species of *Eucalyptus* (Pryor, 1952; Bamber and Humphreys, 1965). However, their combined impact on the growth of eucalypts has rarely been measured. Jacobs (1955), and Penfold and Willis (1961) both remark that eucalypts in Australia do not develop large crowns as rapidly as eucalypts introduced overseas because of large numbers of phytophagous insects which destroy foliage and buds several times per year in Australia but not in the overseas environments.

Two studies have tested the hypothesis that consistently high levels of damage by groups of host specific insect species can affect the species composition of *Eucalyptus* communities. Codominant trees in an association are frequently members of different subgenera of *Eucalyptus* (Pryor, 1959). Burdon and Chilvers (1974a) hypothesized that parasites (i.e., fungi and leaf-eating insects) are restricted to one or other of the different subgenera and that this might influence the species patterns in *Eucalyptus* associations. On four sites, each with two codominant eucalypts; one from each of the subgenera *Symphyomyrtus* and *Monocalyptus*; they measured the damage done by chewing insects and fungi. Of 17 common insect species (i.e., those contributing at least 1% of the total individuals collected), seven were restricted to, and another six showed a strong preference for, one or other of the subgenera. Chewing insects removed from 3 to 38% of the leaf area at the different sites but losses were very similar on both *Eucalyptus* species at any one site. However, damage from fungi was much greater overall and differed markedly between species within a site. They concluded that about 70% of the damage to the *Symphyomyrtus* species at all sites was due to host specific insects and fungi whereas the proportion of host specific damage on *Monocalyptus* species was usually lower and more variable between sites (1, 30, 37, and 83%). The authors concluded that host specificity should be sought primarily at the subgeneric level as they originally hypothesized but their experimental design did not provide an opportunity to look at host preferences among more closely related species on the same site. In the examples I have given in this paper (where sufficient information is provided on host ranges) three insect species were restricted to a single subgenus while six insect species fed on trees in two or more subgenera. Regardless of whether or not they were restricted to particular taxonomic groups, the nine species distinguished between trees within subgenera, sometimes between very closely related species.

Burdon and Chilvers (1974a) compared damage in two adjacent stands with different proportions of *Eucalyptus dalrympleana* and *E. pauciflora*: a dense immature stand with 53% *E. dalrympleana* and a mature stand with 28% *E. dalrympleana*. They found twice as much parasite damage on *E. dalrympleana* as on *E. pauciflora* in the immature stand and about equal damage on the mature trees. They suggested that the amount of parasite attack is density-dependent and contributed to greater thinning of *E. dalrympleana* than *E. pauciflora* in the sapling stage.

In my work (Morrow, in prep.) I have tested the hypothesis that insects with host preferences influence the composition of an alpine woodland association composed of three eucalypt species, *E. perriniana* (subgenus *Symphyomyrtus)* and *Eucalyptus stellulata* and *E. pauciflora* (subgenus *Monocalyptus)*. Of 80 insect species, 63% fed upon only one *Eucalyptus* species, 30% fed on two hosts and of these, two-thirds fed on the two *Monocalyptus* species. Only 8% of the species were generalists that fed on all three *Eucalyptus* species. Of the total individuals collected almost half belonged to taxa specific to a single host while only 3% were in taxa which ate all three eucalypts (Morrow, 1977).

I compared the growth rates of the two *Monocalyptus* species: in the absence of insect attack, *E. stellulata* had a much higher rate of dry matter production than did the dominant species, *E. pauciflora*. At maturity these two species have the same growth habit and are approximately the same size, but, in an even-aged woodland of 35-year old trees, individuals of *E. stellulata* were on an average only one-half the height and diameter of the associated *E. pauciflora*. The apparent suppression of *E. stellulata* may be related to insect attack. Both the number of insect species and individuals were substantially greater on *E. stellulata* than on *E. pauciflora*. In a two-year period, leaf-chewing insects removed 50% of the leaf area and 96% of the shoots of *E. stellulata*, whereas for *E. pauciflora* the corresponding values were 38 and 76%. Furthermore, *E. stellulata* had much greater numbers of sap-sucking insects whose effects were not measured. Experiments using insecticides and examination of growth rings on both sprayed and control trees of both species showed consistently that growth of *E. stellulata* had been suppressed much more by insects than that of *E. pauciflora* for at least 11 years (Morrow and La Marche, in prep.). Thus, in this community differential impact on the productivity of coexisting species appears to prevent competitive exclusion of a slower growing species because its competitor is the preferred host of numerous insect species which severely depress its growth. This less preferred species has a higher net productivity and may ultimately exclude *E. stellulata* unless the canopy is prevented from closing by a major disturbance such as fire.

Discussion

As a group, eucalypts suffer chronic, high levels of insect damage which may impair growth but (except for some very sensitive species) usually does not cause significant mortality. The ability of *Eucalyptus* to tolerate defoliation may be due to its unusual growth characteristics. Leaves can be produced continuously throughout the growing season so trees have tremendous potential to replace damaged foliage rapidly. The amount and rapidity of tissue regeneration is affected by the timing, intensity, and repetition of defoliation. Regeneration becomes less vigorous after starch reserves have been depleted by heavy or repeated attack, or by a period of very rapid growth. In one respect the rapid replacement of leaves may be disadvantageous to the eucalypt—it may provide a ready source of food for certain insects throughout the growing season. Infestation may then become self-perpetuating with continually regenerating tissue available to maintain successive generations of insects at high densities.

The timing of leaf production may vary between trees of the same species even on the same site (Ashton, 1975; pers. obser.). This variation may enable individual trees to escape attack if the timing of leaf flushing differs from that of neighboring trees (L. R. Fox, pers. comm.). Some data suggest that many ovipositing females of several important defoliating species do not move far from where they pupate (Campbell, 1960; Clark, 1962; Carne, 1966; Carne et al., 1974). Consequently, individual trees may escape attack if adjacent, unsynchronized trees have tissues of a more attractive age.

Over its entire geographic range, an insect species may feed on several different *Eucalyptus* species. Some insects may be largely restricted to eucalypt species within a single series (L. R. Fox, pers. comm.) or in a single subgenus (Moore, 1970). On the other hand, many show host preferences that are not related to taxonomic grouping. Even when a species restricts its attack to a particular taxomonic group, it often has clear preferences among the potential hosts. The important point is that within an association, insects may be selective and this can have a significant impact on the dynamics of *Eucalyptus* communities.

For many insects, restriction to a subset of eucalypt species in the field is not necessarily obligatory. Carne (1966), de Little and Madden (1975), and L. R. Fox (pers. comm.) have shown that, for several beetles, growth and reproduction may be as good as or better on coexisting, but less preferred hosts, or even on species never eaten in the field, than on the preferred host. Thus, host preferences may be related to plant chemistry or to such factors as leaf shape (Carne, 1965) or the timing of leaf production (L. R. Fox, pers. comm.).

The strength of insect preference for different hosts within an association may have important consequences for the community. Clark (1962), Carne (1965), and Morrow (in prep.) have found that within an association some insect species overflowed to a less-acceptable species when it was near a preferred host which was saturated with insects. Thus, the spatial pattern of trees may be important. Under some circumstances, it may be advantageous for less acceptable eucalypts to be in monospecific stands where they can escape overflow parasites. This would be especially true for a species such as *E. pauciflora* which has low numbers of host specific insects whose populations are normally small, regardless of whether the tree grows in mixed or monospecific stands (Burdon and Chilvers, 1974a, b; Morrow, in prep.). Conversely, for *E. dalrympleana*, insect attack may increase as relative abundance increases in an association (Burdon and Chilvers, 1974a).

In other parts of the world it has been demonstrated that selective grazing by mammals can either increase or decrease species diversity and abundance of vegetation in pastures and disturbed heathlands (Harper, 1969). In undisturbed forests, selective grazing by deer can reduce and even eliminate preferred forage species when deer populations are high (Klein, 1970). A similar role for insects with host preferences has been suggested because it is known from the literature on biological control that insects can selectively and dramatically alter the abundance of introduced plants which have become weeds (Harper, 1969).

Summary

Eucalyptus species are heavily and chronically attacked by insects. This attack may severely depress plant growth but it does not usually kill the plant. Within an association of *Eucalyptus* species, insects normally feed on only one or a few of the available hosts and this preferential attack may lead to differential impact on the growth of co-existing eucalypt species. Differential grazing may alter the competitive interactions of the plants and thereby affect community species composition and structure.

Acknowledgments. I thank Drs. L. R. Fox and D. C. Potts for reading and constructively commenting on the manuscript.

References

Ashton, D.H.: Seasonal growth of *Eucalyptus regnans* F. Muell. Australian J. Botany 23, 239-252 (1975)

Bamber, R.K., Humphreys, F.R.: Variations in sapwood starch levels in some Australian forest species. Australian Forestry 29, 15-23 (1965)

Beadle, N.C.W.: Soil phosphate and the delimitation of plant communities in Eastern Australia. Ecology 35, 370-375 (1954)

Burdon, J.J., Chilvers, G.A.: Fungal and insect parasites contributing to niche differentiation in mixed species stands of eucalypt saplings. Australian J. Botany 22, 103-114 (1974a)

Burdon J.J., Chilvers, G.A.: Leaf parasites on altitudinal populations of *Eucalyptus pauciflora* Sieb. ex Spreng. Australian J. Botany 22, 265-269 (1974b)

28

Campbell, K.G.: Preliminary studies in population estimation of two
    species of stick insect (Phasmatidae: Phasmatodea) occurring in
    plague numbers in highland forest areas of south-eastern Australia.
    Proc. Linn. Soc. N.S.W. 85 121-141 (1960)
Campbell, K.G.: The biology of *Roeselia lugens* (Walk.), the gum-leaf
    skeletonizer moth, with particular reference to the *Eucalyptus camald-
    ulensis* Dehn (River Red Gum) Forests of the Murray Valley Region.
    Proc. Linn. Soc. N.S.W. 87, 316-338 (1962)
Carne, P.B.: Distribution of the eucalypt-defoliating sawfly, *Perga
    affinis affinis* (Hymenoptera). Australian J. Zool. 13, 593-612 (1965)
Carne, P.B.: Ecological characteristics of the eucalypt-defoliating
    chrysomelid beetle, *Paropsis atomariaol*. Australian J. Zool. 14, 647-672
    (1966)
Carne, P.B., Greaves, R.T.G., McInnes, R.S.: Insect damage to plan-
    tation-grown eucalypts in north coastal N.S.W., with particular
    reference to Christmas beetles (Coleoptera: Scarabaeidae). J.
    Australian Ent. Soc. 13, 189-206 (1974)
Clark, L.R.: The general biology of *Cardiaspina albitextura* (Psyllidae)
    and its abundance in relation to weather and parasitism. Australian
    J. Zool. 10, 537-586 (1962)
Clark, L.R., Dallwitz, M.J.: The life system of *Cardiaspina albitextura*
    (Psyllidae), 1950-74. Australian J. Zool. 23, 523-561 (1975)
Cremer, K.W.: Emergence of *Eucalyptus regnans* seedlings from buried seed.
    Australian Forestry 29, 119-124 (1965)
Cremer, K.W.: Treatment of *Eucalyptus regnans* seed to reduce losses to
    insects after sowing. Australian Forestry 30, 162-174 (1966)
Cremer, K.W.: Effects of partial defoliation and disbudding on height
    growth of *Eucalyptus regnans* saplings. Australian For. Res. 6(1),
    41-42 (1972)
Cremer, K.W.: Ability of *Eucalyptus regnans* and associated evergreen
    hardwoods to recover from cutting or complete defoliation in dif-
    ferent seasons. Australian For. Res. 6(2), 9-22 (1973)
Cremer, K.W., Mount, A.B.: Early stages of plant succession following
    the complete felling and burning of *Eucalyptus regnans* forest in the
    Florentine Valley, Tasmania. Australian J. Botany 13, 303-322 (1965)
Dexter, B.D.: Flooding and regeneragion of red river gum, *Eucalyptus
    camaldulensis* Dehn. For. Comm. Vict. Bull. 20, 35 p, (1967)
Florence, R.G.: The application of ecology to forest management with
    particular reference to eucalypt forests. Proc. Ecol. Soc. Aust-
    ralia 4, 82-100 (1971)
Gilbert, J.M.: Forest succession in the Florentine Valley, Tasmania.
    Royal Soc. Tas. (papers and proc.) 93, 129-152 (1959)
Greaves, R.: Insect defoliation of eucalypt regrowth in the Florentine
    Valley, Tasmania. Appita 19, 119-126 (1966)
Greaves, R.: The influence of insects on the productive capacity of
    Australian forests. Australian For. Res. 3(1), 36-45 (1967)
Grose, R.J.: A study of some factors associated with the natural re-
    generation of alpine ash, *Eucalyptus delegatensis* R.T. Baker, syn *E.
    gigantea* Hook. F. For. Comm. Vict., Bull. 4, (1957)
Grose, R.J., Zimmer, W.J.: 1958. The collection and testing of seed
    from some Victorian eucalypts with results of viability tests.
    For. Comm. Vict., Bull. 10, (1958)
Hadlington, P., Hoschke, F.: Observations on the ecology of the phas-
    matid *Ctenomorphodes tessulata* (Gray). Proc. Linn. Soc. N.S.W. 84, 146-159
    (1959)
Hardin, G.: The competitive exclusion principle. Science 131, 1292-
    1298 (1960)
Harper, J.L.: The role of predation in vegetational diversity. Brook-
    haven Symp. Biol. 22, 48-62 (1969)
Howard, T.M., Ashton, D.H.: Studies of soil seed in snow-gum woodland
    (*Eucalyptus pauciflora* Sieb. ex Spreng var alpina (Benth) Ewart). Vict.
    Nat. 84, 331-335 (1967)

Hutchinson, G.E.: The Ecological Theater and the Evolutionary Play. New Haven: Yale Univ. Press, 1966, 139 p.

Jacobs, M.R.: Growth Habits of the Eucalypts. Canberra: Commonw. Govt., 1955, 262 p.

Kile, G.A.: Insect defoliation in the eucalypt regrowth forests of southern Tasmania. Australian For. Res. 6(3), 9-18 (1974)

Klein, D.R.: 1970. Food selection by North American deer and their response to over-utilization of preferred plant species. In: Animal Populations in Relation to Their Food Resources. Brit. Ecol. Symp. No. 10. Watson, A. (ed.). Oxford and Edinburgh: Blackwell, 1970, pp. 25-46

Larsen, E.: Germination of *Eucalyptus* seed. For. and Timber Bureau Leaflet No. 94. 1965

de Little, D.W., Madden, J.L.: Host preference in the Tasmanian eucalypt defoliating paropsini (Coleoptera: Chrysomelidae) with particular reference to *Chrysophtharta bimaculata* (Oliver) and *C. agricola* (Chapuis). J. Australian Ent. Soc. 14, 387-394 (1975)

Mazanec, Z.: The effect of defoliation by *Didymuria violescens* (Phasmatidae) on the growth of alpine ash. Australian Forestry 30, 125-130 (1966)

Mazanec, Z.: Mortality and diameter growth in mountain ash defoliated by phasmatids. Australian Forestry 31, 221-223 (1967)

Mazanec, Z.: Influence of the jarrah leaf miner on the growth of jarrah. Australian Forestry 37, 32-42 (1974)

Moore, C.W.E.: Interaction of species and soil in relation to the distribution of eucalypts. Ecology 40, 734-735 (1959)

Moore, K.M.: Observations on some Australian forest insects. 4. *Xyleborus truncatus* Erichson 1842 (Coleoptera: Scolytidae) associated with dying *Eucalyptus saligna* Smith (Sydney Blue-Gum). Proc. Linn. Soc. N.S.W. 84, 186-193 (1959)

Moore, K.M.: Observations on some Australian forest insects. 24. Results from a study of the genus *Glycaspis* (Homoptera: Psyllidae). Australian Zoologist 15, 343-376 (1970)

Morrow, P.A.: Host specificity of insects in a community of three co-dominant *Eucalyptus* species. Australian J. Ecol. (1977, In press)

Parsons, R.F., Specht, R.L.: Lime chlorosis and other factors affecting the distribution of *Eucalyptus* on coastal sands in southern Australia. Australian J. Botany 15, 95-105 (1966)

Penfold, A.R., Willis, J.L.: The Eucalypts. New York: Interscience, 1961, 551 p.

Pook, E.W., Moore, C.W.E.: The influence of aspect on the composition and structure of dry sclerophyllous forest on Black Mountain, Canberra, A.C.T. Australian J. Botany 14, 223-242 (1966)

Pryor, L.D.: Variable resistance of leaf-eating insects in some eucalypts. Proc. Linn. Soc. N.S.W. 77, 364-368 (1952)

Pryor, L.D.: Genetic control in *Eucalyptus* distribution. Proc. Linn. Soc. N.S.W. 78, 8-18 (1953)

Pryor, L.D.: Species distribution and association in *Eucalyptus*. In Biogeography and Ecology in Australia. Keast, A., Crocker, R.L., Christian, C.S. (eds.). The Hague: Junk, 1959, pp. 461-471

Pryor, L.D., Johnson, L.A.S.: A Classification of the Eucalypts. Canberra: Australian Nat. Univ., 1971, 102 p.

Readshaw, J.L., Mazanec, Z.: Use of growth rings to determine past phasmatid defoliations of alpine ash forest. Australian Forestry 33, 29-36 (1969)

Shepherd, K.R.: Defoliation of alpine ash, *E. delegatensis*, by Phasmids. For. Comm. N.S.W. (1957)

Wallace, M.M.H.: The biology of the jarrah leaf miner, *Perthida glyphopa* Common (Lepidoptera: Incurvariidae). Australian J. Zool. 18, 91-104 (1970)

Winterhalter, E.K.: Differential resistance of two species of *Eucalyptus* to toxic soil manganese levels. Australian J. Sci. 25, 363 (1963)

Chapter 3

Resource Utilization by Colonial Lepidoptera Defoliators

J. T. CALLAHAN

In the southeastern United States there are indigenous populations of
two colonial Lepidoptera: the eastern tent caterpillar, *Malacosoma americanum* (Lasiocampidae); and the fall webworm, *Hyphantria cunea* (Arctiidae).
The fall webworm has two races which are easily distinguishable by their
larval coloration (orange and black) and their selection of plant hosts
(Oliver, 1964). Both *M. americanum* and *H. cunea* build silken nests or tents
around branches and foilage of trees. Larvae congregate within the
central, denser portions of the tents when resting.

Incidence of Colonies on Hosts

Over a five-year-period I examined almost 7000 colonies of the three
types of colonial larvae. Each type of colony exhibited rather distinct
host preferences (Table 1). The three types of larvae utilized a total
of 28 plant species including several trees, shrubs, and vines. The
orange fall webworm had the broadest range of hosts (21), followed by
the black webworm (15), and the eastern tent caterpillar (12). Oliver's
(1964) data for Louisiana indicated that the black race webworm had a
broader range of hosts than the orange race. Warren and Tadic (1970)
reported that at least 636 plant species of 326 genera were acceptable as food by both races of the webworm. No such complete host information was found for the eastern tent caterpillar.

Table 1. Frequency of natural occurrence of colonial Lepidoptera
larvae on different plant species in the Savannah River Plant Reserve

Larvae	Host species	Number colonies	Percent colonies
Eastern tent caterpillar	*Prunus serotina*	2511	86.5
	Sassafras albidum	146	5.0
	All others (10)	247	8.5
	Total	2904	100.0
Orange fall webworm	*Diospyros virginiana*	2372	95.0
	All others (20)	125	5.0
	Total	2497	100.0
Black fall webworm	*Salix caroliniana*	587	37.0
	Liquidambar styraciflua	431	27.2
	Liriodendron tulipifera	174	11.0
	Nyssa sylvatica	163	10.3
	Diospyros virginiana	57	3.6
	Carya tomentosa	46	2.9
	All others (9)	129	8.0
	Total	1587	100.0

The study area was the 300-mile^2 Savannah River Plant Reserve, a U.S. Energy Research and Development Administration facility in South Carolina. Forest manipulation within this reserve has been limited to planting, maintenance, and harvest of large pine plantations under the supervision of the U.S. Forest Service. There have been no control programs for either the tent caterpillar or the webworm since the closing of the reserve in the early 1950s. It is thus safe to assume that the insect population levels are representative for southeastern mixed forest types and managed pine plantations.

Habitat and Host Selection

The three types of larvae occur on different subsets of hosts. The distinctions among those subsets become even more striking when the habitats are examined physically. The principal hosts of the eastern tent caterpillar, Carolina cherry, *Prunus serotina*, and sassafras, *Sassafras albidum*, occur mainly in relatively open areas along abandoned fence rows and in the wide rights-of-way of highways, railroads, and electrical transmission lines. The persimmon, *Diospyros virginiana*, main host of the orange webworm, also occurs along such rights-of-way, but mainly on the edges adjacent to mature forest and along the unpaved roads which border pine plantations. The plant species selected by the black webworm occur in mature hardwood stands on mesic sites where willows, *Salix caroliniana*, inhabit stream banks and the ubiquitous persimmons inhabit the edges.

Apparently, these insects (at endemic densities at least) are inhabitants of edges—edges of stream channels, roadways, and old fields. A major factor contributing to this characteristic is the females' selection of open flyways when seeking oviposition sites. Of 60 gravid females released for observation (25 eastern tent caterpillar, 20 orange fall webworm, and 15 black fall webworm) in appropriate habitats, none flew farther than 100 m before alighting, and all stayed within the open spaces. The females' predilection for open habitats may account for the fact that only 5% of the colonies of any of the three types were found more than ten meters from an open flyway. This characteristic has also been reported by other investigators (Morris and Bennett, 1967; Ito et al., 1970).

Since more than 95% of all colonies were found within ten meters of the closest open area, the preferred habitat for these defoliators is actually long, narrow strips adjacent to forest openings.

Estimating Food Resources and Their Consumption

The first step in assessing the available resource base for these defoliators consisted of delineating their preferred habitat. Since females are unlikely to venture far from an open flyway in selecting an oviposition site, the resources are limited to those hosts within or close to forest openings. Preferred host plants which grow in areas not immediately adjacent to or within a flyway comprise unavailable resources—at least at endemic densities.

The second step consisted of measuring the density and size (d.b.h.) of the total suite of preferred hosts within the previously defined habitats for the three larval types and then estimating available

leaf biomass from stem diameter (e.g. Cotter and Monk, 1967). The final
step consisted of converting leaf biomass to caloric equivalents
(Phillipson, 1964) as shown in Table 2.

Table 2. Resource base available in specific habitats of three colonial
larvae

Larvae	Host species	Density stems/ha	dbh (cm)	Resource base 10^3 Kcal/ha
Eastern tent	*Prunus serotina*	51.4	6.1	66
caterpillar	*Sassafras albidum*	148.9	1.4	31
	Total	--	--	97
Orange fall webworm	*Diospyros virginiana*	99.4	3.4	66
Black	*Salix caroliniana*	47.8	2.0	14
fall webworm	*Liquidamber styraciflua*	8.9	7.3	14
	Liriodendron tulipifera	3.7	15.3	14
	Nyssa sylvatica	12.1	7.6	17
	Diospyros virginiana	18.1	5.8	21
	Carya tomentosa	0.9	21.6	4
	Total	--	--	84

Laboratory and field studies (Callahan, 1972) have shown that the two
yearly generations of the orange fall webworm consume approximately
3200 kcal of leaf material per hectare in any given year. This amounts
to about 4.8% of the leaf material available on the principal host,
persimmon, within its characteristic habitat (Table 3). The black
webworm and the eastern tent caterpillar consume 2.3% and 3.4% respectively, of the leaf resources in their habitats. It is apparent that
the total population of colonial Lepidoptera consume annually less than
5% of their preferred resource base, and that their proportional impact on total annual leaf production in these ecosystems would be
much less.

Potential for Defoliation

Since these insects consume only about five percent of their preferred
resource base, one might conclude that they are insignificant in their
respective ecosystems.

However, it is enlightening to examine the reproductive rates of these
consumers. Natural controls on these populations are normally efficient, but what happens if those mechanisms break down for even one
annual cycle? To illustrate the outcome of such a breakdown, I calculated maximum rates of increase (r_{max}) from the mean cohort size, sex
ratios, and number of annual generations (Table 4). If survival were
100% for only one year, then the number of insects in the next generation would require 600-100,000-fold more foliage than is available
from the preferred hosts. Under these circumstances, the insect populations could easily consume all of the available leaf biomass in
their ecosystems.

Table 3. Estimated consumption of resources by three colonial larvae

Larvae	Generations	Colonies	Consumption	Resources	Resource consumption
	no/yr	no/ha/yr	Kcal/ha/yr	Kcal/ha/yr	Percent
Eastern tent caterpillar	1	19.4	3300	97,000	3.4
Orange fall webworm	2	13.7 (9.5 + 4.2)	3200	66,000	4.8
Black fall webworm	3	8.2 (2.1 + 4.3 + 1.8)	1900	84,000	2.3

Table 4. Potential energy consumption by three colonial larvae after one year of maximal population increase

Larvae	Rate of increase	Potential colonies	Potential consumption	Potential consumption/ available resources
	r_{max}/yr	no/ha/yr	Kcal/ha/yr	
Eastern tent caterpillar	3.4	3.3×10^3	5.6×10^5	5.8×10^2
Orange fall webworm	11.1	6.5×10^5	1.5×10^8	2.3×10^3
Black fall webworm	15.5	3.0×10^7	7.0×10^9	8.3×10^4

34

Summary

Larval colonies of the eastern tent caterpillar and two "races" of the fall webworm are present in southeastern forests. Their mean annual densities are normally low, ranging from 8.2 to 19.4 colonies/ha. The three types are highly selective among potential host plant species at endemic densities, and their preferred habitats can be differentiated on that basis. Consumption of the resource base under conditions of endemic density does not exceed 5% of that available from the small suite of preferred host plants. Maximum rates of population increase for these defoliators are high, from 3.4 to 15.5. Natural controls on these populations normally function efficiently, but if they were to be removed or break down for as little as one annual cycle the colonial Lepidoptera populations could easily consume all leaf material in their native forest ecosystems.

References

Callahan, J.T.: Ecological energetics and population dynamics of the fall webworms, *Hyphantria cunea* Drury (Lepidoptera: Arctiidae). Ph. D. Dissertation, Univ. Georgia, 1972, 119 p
Cotter, D.J., Monk, C.D.: 1967. Analysis of leaf-stem relationships for eighteen tree species. Sav. River. Ecol. Lab., Ann Rep. 1967, 18-19
Ito, Y., Shibazake, A., Iwahashi, O.: Biology of *Hyphantria cunea* Drury (Lepidoptera: Arctiidae) in Japan. XI. Results of road survey. Appl. Ent. Zool. 5(3), 133-144 (1970)
Morris, R.F., Bennett, C.W.: Seasonal population trends and extensive census methods for *Hyphantria cunea*. Can. Entomologist 99, 9-17 (1967)
Oliver, A.D.: A behavioral study of two races of the fall webworm in Louisiana. Ann. Ent. Soc. Am. 57, 192-194 (1964)
Phillipson, J.: A miniature bomb calorimeter for small biological samples. Oikos 15, 130-139 (1964)
Warren, L.O., Tadic, M.: The fall webworm. Univ. Arkansas. Div. Agr. Bull. 759, 106 p (1970)

Chapter 4

Species Structure of Bumblebee Communities in North America and Europe

D. INOUYE

The competitive exclusion principle states that two species cannot coexist on the same limiting resource. It is possible to view bumble-bees *Bombus* spp. (and other pollinators) as a limiting resource which is partitioned by the flowering plants in a community. I have chosen the alternative view which considers flowering plants as a limiting resource that is partitioned by bumblebees. The fact that bumblebees are annual species, while many of the plants they visit are long-lived perennials, lends support to this interpretation. Furthermore, many plant species may have other means of effecting pollination in addi-tion to bumblebees. Because the relationship between bumblebees and flowers is generally mutualistic, consideration of both views is likely to be most fruitful.

Methods

I have studied the resource partitioning, diversity, distribution, and abundance of bumblebee species in the Colorado Rocky Mountains. Eleven species of *Bombus* occur in the East River Valley from an elevation of 2740 m to 3267 m. Not all of these species are equally abundant at all sites in the valley, however. Individual species are usually associ-ated with particular altitudinal ranges in the valley, and may occur only rarely above or below these altitudes. In the vicinity of the Rocky Mountain Biological Laboratory (Gothic), at an elevation of 2896 m (9500 ft), there have been three or four common species of bumblebees. In Virginia Basin, at an elevation of 3505 m (11,500 ft), there have been three common species. Although only one species is common to both sites, there are other similarities between the bumble-bees present in the two areas.

Biologists commonly study differences in resource utilization by exam-ining morphological characteristics of a species that indicate the position of its utilization curve on the appropriate resource dimen-sion (Hespenheide, 1973). The appropriate morphological character in bumblebee is proboscis length, which is related to the size of flowers which the bumblebee species visits. We can then consider the range of proboscis lengths exhibited by bumblebees in Gothic and Virginia Basin.

Results and Discussion

Community Structure in Colorado

It is apparent that, with one exception, there is one species repre-sentating each of three length classes (Fig. 1). The ratios between proboscis lengths of adjacent species fall (in all but one case) in the range of 1.2 to 1.4, in accordance with Hutchinson's (1957) ob-

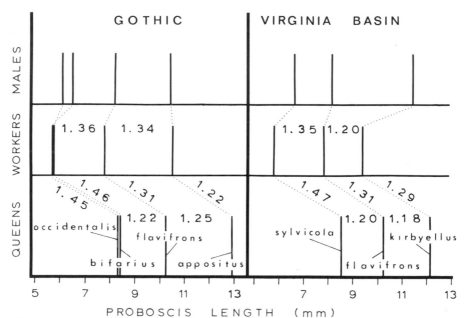

Fig. 1. Proboscis lengths of common bumblebee species present in Gothic (elevation 2896 m) and Virginia Basin (elevation 3505 m). The numbers between species and castes are the ratio of the larger over the smaller measurements of two proboscides. *Bombus occidentalis* is a nectar robber

servation that in certain groups of animals, otherwise similar species differed in the size of their feeding apparatus by a constant factor of 1.2 to 1.4. This observation suggests that niches are regularly and widely spaced and further implies that the pattern of resource utilization among bumblebee species results from competition. Assuming that proboscis length is a good indicator of resource utilization, one can infer that species adjacent on the size scale also differ in mean food size by a constant ratio.

Evidence for competition between bumblebees of similar proboscis length is provided by reports that species of similar proboscis length often replace each other geographically or vertically. For example, Løken (1950) described the apparent replacement of *Bombus agrorum* by *Bombus balteatus* over an altitudinal gradient. Laidlaw (1930) mentioned several cases of replacement of one species by another with increasing latitude. Hanninen (1962) confirmed earlier observations that *Bombus lapidarius* and *B. agrorum*, which have similar proboscis lengths, are found "to some degree in inverse proportion." He found that *B. lapidarious* was most numerous in the only trial in which *B. agrorum* was not present. Hänninen also reported that *Bombus lucorum* and *Bombus jonellus* (both nectar robbers) replace each other from north to south. Peters (1967) cited two cases of apparent species replacement but one involved bees of different proboscis lengths. Koeman-Kwak (1973) also reported an apparent inverse correlation of the abundance of two species.

The single case in this study in which a proboscis length is represented by two species *(Bombus bifarius* and *Bombus occidentalis)* provides the exception that proves the rule. *B. bifarius* and *B. occidentalis* were both common species in Gothic in 1974. *B. occidentalis* is a nectar robber, which means that it collects nectar primarily by biting through the corolla tubes of flowers to gain access to the nectar (Proctor and Yeo, 1973). This behavior is characteristic of certain species of *Bombus*, all of which have short proboscis length. Several species from North America and Europe commonly exhibit this behavior. By employing this behavior, *B. occidentalis* is able to collect nectar from flowers from which it would normally be excluded, including flowers not visited by *B. bifarius*, and even flowers not visited by any other bumblebee species. Thus, *B. occidentalis* does not compete strongly with other bee species of short proboscis length for the same floral resources.

Community Structure in Europe and North America

The fact that the total group of bumblebee species in the East River Valley falls into three classes of proboscis length, and that each class was represented by a single species of legitimate forager in my two study sites, suggested that there might be a common pattern to bumblebee communities elsewhere. Unfortunately, it is not possible to determine from the literature, most of which is from Europe, how common this pattern is. There has been no authoritative study of proboscis lengths of European bumblebee species, and most studies of bumblebees either considered extremely large areas or failed to indicate which species were represented by all three castes. Brian's (1957) study of bumblebees in the area of the Firth of Clyde in Scotland presents an interesting comparison, however (Fig. 2).

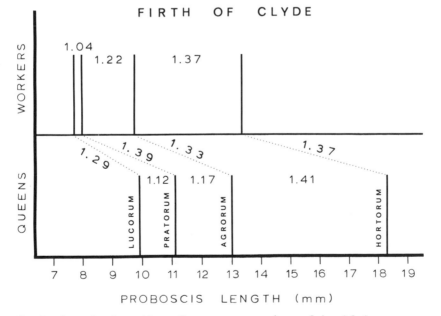

Figure 2. Proboscis lengths of common species of bumblebees present on the north bank of the Firth of Clyde, Scotland. *Bombus lucorum* is a nectar robber. The numbers between species and castes are the ratio of the larger over the smaller measurements of two proboscides (data from Brian, 1957)

Brian considered four species to be important in her study area. One
of these, *B. lucorum*, was a nectar-robbing species, and the other
three represented long, intermediate, and short proboscis length
classes. These classes were again separated by ratios of 1.2 to 1.4.
However, on an absolute scale the size classes do not correspond directly to those found in Colorado. In fact, the short and intermediate proboscis lengths in Scotland correspond to the intermediate
and long lengths respectively, in Colorado. This difference appears
to be consistent throughout Europe. There are no species of *Bombus*
in Europe with proboscis lengths as short as those found in North
American species of *Bombus*. The proboscis length of honeybee, *Apis
mellifera*, workers, however, corresponds closely to that of the short
length class in North America. Honeybees are native to Europe, but
not the New World.

D. Morse (pers. comm.) found four common bumblebee species present
in a study area of Maine, representing the same pattern of proboscis
lengths and foraging behaviors already described for Gothic. Stephen
(1955) reported the same four species, with a smaller number of a
fifth species, from alfalfa fields in southeastern Manitoba. In
Aberdeen, Scotland, Laidlaw (1930) found three "really common" species
and two additional species that were "also abundant but less so than
the first three." Of the three common species one represents the long
proboscis class, and the other two represent the short class of which
one is a nectar robber. The remaining two species both represent the
intermediate class, but one of these was only "locally" abundant. G.
Pyke (pers. comm.) found three common species present in parts of
Utah, representing the three proboscis length classes. In some areas,
however, not all classes were represented. A. Løken (pers. comm.) has
found that *Bombus* species belonging to the same proboscis size class
often coexist, although she does not state whether all castes were
present.

The number of species of bumblebees in plant communities is not fixed.
For example in Colorado there were three common species (four if a
robber were present); while in other areas there were fewer or more
than this number. Several factors might determine the number of species
which can coexist:

1. If a portion of the food resource continuum of corolla tube length
is not represented in a community, we might expect to find that the
proboscis length class which would normally utilize that portion of
the resource spectrum is missing.

2. If flowering during the summer is not continuous, or if abundances
drop markedly at some point, we might expect to find fewer than three
species present.

3. If the floral resources are in fact so abundant that there is no
competition for food, more than three (or four) species may stably
coexist in a community.

To what extent these factors can explain the observed differences in
species abundance is as yet unknown.

Plant-Bumblebee Community Structure

The relationship between bumblebees and the plants they pollinate is
mutualistic, and often obligatory. Considerations of the interactions
between these two groups of organisms should prove more fruitful than

isolated studies of either group. Toward this end King et al. (1975) have developed a model to examine relationships between species diversity of the two groups. They restricted their model to self-incompatible plants that are totally dependent upon animals for pollen transfer, and to animals, such as bumblebees, that received all of their sustenance from flowers at some point in their life cycle. They proceeded to demonstrate that equilibrium numbers of plant and pollinator species may be established. They also found that while the diversity of the equilibrium assemblages is unique, the species composition is not. The evidence from Gothic, Virginia Basin, and other areas supports this conclusion.

Given that there appears to be an equilibrium number of bumblebee species present in a community, we may infer that there is a similar number of bumblebee pollinated plant species present in areas with similar numbers of bumblebee species. Data are not yet available to test this hypothesis in detail. Moldenke (1975), however, found that the number of food plants per flower visitor (not restricted to bumblebees) was similar in most of the communities he studied. If each bumblebee species requires a certain number of flower resources which it does not have to share with other bumblebee species, the number of bee and flower species present should be positively correlated. The life cycle of a bumblebee colony is generally much longer than the flowering period of a single plant species. Thus, an equilibrium number of flower species in a community should be greater than the corresponding number of bee species.

It is possible to predict a priori some attributes of the plant community one could expect to find associated with a particular group of bumblebees. For example, there is a general correspondence between the proboscis length of a bee species and the corolla tube lengths of the flowers it visits for nectar. So, we can predict that if representatives of three proboscis length classes are present in a bumblebee community, a continuum of flower corolla tube lengths should also be present. Furthermore, there should be corolla tube lengths appropriate for the different proboscis lengths represented by queen and worker castes of each species.

The bumblebee species primarily associated with forest habitats can perhaps be delimited as a subset of the total group. Frison (cited in Fye and Medler, 1954) noted that bumblebees follow the floral succession. The first bee species to emerge from hibernation are those which nest in the woods, where the majority of the flowering plants are at that time. Later emerging species nest in the open, where most of the flowering is occurring at that time.

Summary

The relationship between the reproductive biologies of plants and plant pollinators is the result of a mutualistic evolutionary process generally referred to as coevolution. Evidence indicates that the evolutionary relationships between bumblebee species are well defined, having resulted in separations between coexisting species along the dimension of proboscis length. Given the similarities found between North American and European bumblebees, we can infer that there may also be similarities in the plant communities present on the two continents. First, we can expect to find similar numbers of bumblebee-pollinated plants in communities with similar numbers of bumblebee' species. Second, we can predict to some degree the corolla tube sizes of those flowers. What remains now is to test some of these ideas.

References

Brian, A.D.: Differences in the flowers visited by four species of bumblebees and their causes. J. Animal, Ecol. 26, 71-98 (1957)

Fye, R.E., Medler, J.T.: Spring emergence and floral hosts of Wisconsin bumblebees. Wisc. Acad. Sci., Arts Lett. 43, 75-82 (1954)

Hänninen, P.: Bumblebee species on red clover in Central Finland. Pub. Finnish State Agric. Res. Board #197, 1962

Hespenheide, H.A. Ecological inferences from morphological data. Ann. Rev. Ecol. Syst. 4, 213-229 (1973)

Hutchinson, G.E.: Concluding remarks. Cold Spring Harbor Symp. Quant. Biol. 22, 415-427 (1957)

King, C.E., Gallaher, E.E., Levin, D.A.: Equilibrium diversity in plant-pollinator systems. J. Theoret. Biol. 53, 263-275 (1975)

Koeman-Kwak, M.: The pollination of *Pedicularis palustris* by nectar thieves (short-tongued bumblebees). Acta Botan. Neerl. 22, 608-615 (1973)

Laidlaw, W.B.R.: Notes on some bumblebees and wasps. Scottish Naturalist 184, 121-125 (1930)

Løken, A.: Bumble bees in relation to *Aconitum septentrionale* in Western Norway. Nor. Entomol. Tidsskr. 8, 1-16 (1950)

Moldenke, A.R.: Niche specialization and species diversity along a California transect. Oecologia 21, 219-242 (1975)

Peters, G.: Ein Frühsommeraspekt der Hummelfauna von Monchgut auf Rügen. Deut. Entomol. Z. 14, 125-137 (1967)

Proctor, M., Yeo, P.: The pollination of Flowers. London: Collins 1973, 418 p.

Stephen, W.P.: Alfalfa pollination in Manitoba. J. Econ. Ent. 48, 543-548 (1955)

Chapter 5

Pollination Energetics: An Ecosystem Approach

B. HEINRICH

Introduction

Classical pollination studies focused on flower morphology relative
to morphology and behavior of pollinators (Müller, 1883; Knuth,
1906-9). More recent studies have emphasized (1) pollinator foraging
behavior as a key to understanding inter-plant gene flow (Levin,
1972a), (2) foraging energetics as a unifying approach to pollination
ecology (Heinrich and Raven, 1972), and (3) ecosystem perspectives
(Pojar, 1974; Stiles, 1975). Unlike morphological coadaptations between flowers and pollinators, the "fit" between caloric food rewards
of flowers and energy expenditures of their pollinators is often not
readily apparent, except grossly. However, some precision has been
demonstrated using time-energy budgets in territorial species (Gill
and Wolf, 1975). Apparent "sloppiness" of energy fit may result from
viewing pollinator-plant interactions on a one-to-one basis rather
than from an ecosystem perspective. In addition, many flowers are
utilized by foragers of both high and low energy expenditure. In this
paper I will examine some indirect interactions between plants and
between pollinators at the ecosystem level. Such interactions may
reveal a more precise "fit" than was first apparent.

Plant Interdependence

Most plants provide sustenance to a large variety of animals. In fact,
hundreds of forager species may visit a given flower species (Knuth,
1906-9), although only a few are the "actual" pollinators. It may
appear, therefore, that the nonpollinating flower visitors are minor
parasites. As already pointed out by Baker et al. (1971), a web of
indirect interactions can exist between plants with visitors in common.
These authors observed that almost all "bat flowers" are visited also
by birds, as well as by insects. In the thorn scrub of western Mexico,
for example, bat-pollinated *Ceiba acuminata* (Bombacaceae) is visited
by hummingbirds, various kinds of bees, and other insects that are
irrelevant to the pollination of these plants. However, *Ceiba* may
benefit indirectly by feeding the nonpollinating foragers, for these
foragers are maintained in the ecosystem and pollinate plants which
bats feed on in the wet season when *Ceiba* is no longer in bloom.

Possibly similar plant interdependence has been observed by Wolf
(1970) and Stiles (1975) in hummingbird-pollinated flowers. The birds,
however, also feed on insects. When not nesting the birds can move to
new areas if nectar becomes scarce. Social bees, on the other hand, do
not have that option, and thus they are obligately tied to the local
plant community. Interdependence of bog plants and bumblebee pollinators, for example, has been discussed by various authors (Judd, 1958;
Pojar, 1974; Heinrich, 1975; Reader, 1975; Small, 1976; Heinrich,
1976b). Plants of temperate bogs flower in a steady progression from

spring until late summer. The bumblebees in the bog usually rely on only 2-3 plant species for most of their food at any one time. The absence of flowers for a short time in the summer could result in death of the bee colonies and hence their absence from the habitat for the entire year. As a consequence there is apparent species interdependence of the bumblebee-pollinated plants, since each of the different plant species provides links in the temporal chain of food necessary for the bee's continued survival.

Snow (1965) has made similar inferences about plant-plant interdependence on the basis of seed dispersal by frugivorous birds. He observed a sequential ripening of fruits of 19 species of *Miconia* (Melastomaceae) throughout the year in the Arima Valley of Trinidad. At least two species were fruiting every month of the year, allowing frugivorous birds to remain in the habitat and disperse seeds of all the competing plant species. Unique fruiting and flowering times not only have positive selective value to the individual plants, but also to the other members of the plant community (its temporal neighbors) that compete for pollinators and seed dispersers.

Temporal plant-plant interdependence through pollinator sharing presumably evolved from competition for pollinators (Mosquin, 1971, and others). Levin (1972b) has found experimental evidence showing that butterflies discriminate against the rarer morph of *Phlox drummondii*, suggesting that the dominant flowering plant has an advantage in attracting pollinators. As discussed previously (Heinrich and Raven, 1972), the selective pressure arising from such pollinator behavior should tend to produce synchronized but staggered blooms among competing species.

Recently some aspects of natural selection at the community level, as exerted through indirect effects, have been examined through modeling (Wilson, 1976). Every effect of a species on others in the community loops back on itself. But this feedback is traditionally considered to be neutral inasmuch as the densities of the affected species are temporally and spatially assumed to be the same. In other words, natural selection cannot act differentially to amplify the components of specific feedback loops unless they are isolated. However, Wilson (1976) proposes that evolution at the community level can occur since there usually is, infact, variation in community composition so that organisms would feel their own effects on the community differentially. This framework of thought provides the means to explain evolution at the community level without violating the principle of individual selection. Pollination systems should provide excellent opportunity and case studies for the study of community evolution, since the composition of flowers in them varies spatially, temporally, or both.

Pollinator Interdependence and Pollination

For maximum interplant movement of foragers at minimum energetic cost to plants, the amount of food provided by flowers should be sufficient to attract foragers, yet low enough to keep them moving from one plant to another (Heinrich and Raven, 1972). Carpenter (1976) has recently shown in the field that there is indeed an optimum amount of nectar reward in *Metrosideros collina* (Myrtaceae) flowers, above and below which seed-set is reduced. Low food-reward flowers promote forager vagility, but they can only attract pollinators with low energy demands. On the other hand, high food-reward flowers could feed both high- and low-

energy foragers, but the low energy foragers feeding at high food
reward flowers might be too site-specific for effective cross-pollination.

When different forager species having diverse energy demands utilize
the same flowers, it is not always apparent which are the actual
pollinators. Food removal by one affects the movements, and hence the
impact of the others. For example, the food rewards of a number of
plants in Maine are sufficiently ample to retain small wild bees,
rather than forcing them to move between plants, as is required for
efficient cross-pollination. However, the large food rewards of flowers are usually harvested by bumblebees and the "left-overs" are still
sufficient to attract small bees that generally visit the same flowers
later in the day when bumblebees have ceased utilizing the flowers
(Heinrich, 1976c). Thus, because of the bumblebee's harvests, the small
"solitary bees" visit many more flowers than they would without bumblebees, thereby increasing their potential as pollination agents. Butler
(1945) and Free and Spencer-Booth (1963) have previously observed that
honeybees begin to wander between trees, and thus become effective
pollinators in orchards, only when the food resources are depleted.

Another possible combination of low- and high-energy pollinators on
multi-flowered plants (e.g. trees) is that where the high-energy
foragers serve as inter-plant pollinators, the site-specific low-energy
foragers serve as intra-plant pollinators. Such a system could operate
in the widely dispersed forest trees of the lowland tropics (Bawa,
1974). Massively flowering trees are visited by large numbers of bees
(Frankie and Baker, 1974), some of which (Euglossines) are capable of
long distance movements to and from their nests (Janzen, 1971). The
normal as well as the maximum foraging distances of these bees are
unknown. Another example of large vagile foragers (birds) and small
possibly site-specific foragers (insects) have been examined in Hawaii
by Carpenter (1976). *Metrodiseros collina* has typical "bird flowers"
which do not produce seed when enclosed in fine-mesh bags that exclude
both insects and birds. However, the flowers do produce seeds when
insects have access to the flowers but birds are excluded (Carpenter,
1976). Insects are thus important in the pollination of these "bird
flowers." However, since the trees have many flowers and the flowers
have much nectar, it can be inferred that the insects would be relatively site-specific to individual trees. The birds do not restrict
themselves to individual trees. It is not yet known, however, whether
the inter-plant pollen flow is accomplished solely by the birds and
the insects spread pollen brought by the birds.

Flower Morphology

Selective pressures for shifts in blooming time of co-existing plants
would presumably vary in time and space, depending on neighbors. This
raises the possibility for the evolution of specific sequencing in the
blooming of flowers of different morphologies. Snow (1971) has already
pointed out the existence of temporal sequential ripening of different
types of fruits. Late-maturing fruit tend to be more nutritious, and
well adapted for wintering birds. Similarly, as inferred from the
foraging behavior of pollinators, there should be selective pressure
for similar flowers to bloom in sequence. Even the highly polylectic
bumblebees, as individuals, become highly flower-constant for as long
as their host plants remain remunerative (Heinrich, 1976a). In the
polylectic bees, where an individual's choice leading to flower-constancy is based on food suitability, the initial conditioning to one

species of flower presumably enhances subsequent sampling and conditioning to other flowers of similar appearance. The bee's conservative behavior should select for serial sequencing of blooming in different species of flowers with similar appearance. Furthermore, the second flowers that bloom in a sequence of similar flower species could afford to provide less food reward than the first. Some recent evidence is consistent with this hypothesis: in eight pairs of Maine plant species having flowers of similar appearance, the second to bloom in seven of these pairs provided on the average 74% less nectar than the first (Heinrich, 1975). Variety of flowers at any one time promotes forager constancy. Whether innate or conditioned, the flower constancy of foragers is ultimately limited by the amount of difference they perceive between the flowers of simultaneously blooming plant species. The more that different species resemble one another, the more likely it is that the foragers will visit more than one species, collecting nectar and pollen but not necessarily pollinating the plants. Conversely, the greater the difference among concurrently blooming species, the more likely it is that the foragers will refine their responses (i.e., specialize) and become flower-constant. "Variety" in concurrently blooming flowers presumably has functional significance. At the present time, however, there are no quantitative data that can be used to measure the evolution of flower "variety" at the ecosystem level. MacSwain et al. (1973) observed great variety of petal shape and color among bee-pollinated *Clarkia* in the western United States. The significance of the specific signalling features of these flowers may be to promote flower-constancy of individual foragers, and possibly different species. As shown experimentally by Levin and Berube (1972), differences in flower morphology greatly reduce interspecific pollen transfer, thus enhancing pollination efficiency. Since some hybrids are sterile (Lewis, 1961), bees that stray between species could remove plants from the reproducing population. This should be a large penalty least likely to accrue to those plants with flowers different (both with regard to signalling features as well as functional morphology) from the potential hybridizers.

Differences in the flowers of concurrently blooming plants result not only from signalling cues, but also from different attractants and from flower morphology requiring different manipulative behavior. The flowers may diverge sufficiently to promote species, as well as individual foraging specialists. For example, Stiles (1975) has shown a variety of morphological and other isolating mechanisms among a guild of *Heliconia* species in Costa Rica for hummingbird pollinators.

Adaptations that promote forager-constancy in concurrently blooming plants may be varied and complex. As shown by Levin (1972b) there is advantage for uniformity of floral signals within the species; variants are selected against by pollinators. Rare species, like variants within species, are thus at a disadvantage unless they provide superior attraction. Several examples have suggested that floral mimicry by rare species of common ones may partially compensate for rareness to promote pollination success (Macior, 1974).

Flowers which provide only pollen or which (like many orchids) rely on scent, pseudo-copulation or other "deceit" for pollination, cannot exist in a habitat that does not have other flowers concurrently providing nectar as an energy source. For example, the orchid, *Calopogon tuberosus*, (which has no nectar) almost always blooms together with *Pogonia ophioglossoides* (which has nectar). The flowers of the two species are similar in size, and both have perianths with yellow-white brushes that exhibit strong ultraviolet absorption (Thien and Marcks, 1972). The flowers of both species are pollinated by bees that occasionally visit the two types of flowers sequentially (Thien and Marcks, 1972;

Heinrich, 1975). The flowers differ in that each deposits it pol-
linia on different parts of the bee's body. Because *P. ophioglossoides*
has nectar, *C. tuberosus* need not produce any because it utilizes in-
experienced bees that mistake it for similar nectar producing flowers. The bees probably use color and morphology in flower recognition.
It is of interest, therefore, that the color of *Calopogon* is not
constant but varies from white to violet. Thus bees would have to
learn to *avoid* the white, pink as well as purple flowers in order to
avoid *Calopogon*. Thus flower variety in this nonrewarding plant might
prolong the bee's service to it. Bees also use scent for short-range
flower recognition, but the nonrewarding *Calopogon* flowers are un-
scented (Thien and Marcks, 1972). If *Calpogon* flowers were scented, as
well as of uniform color, bees would presumably learn to avoid them
sooner. I have observed up to 17 consecutive visits to the unrewarding
flowers by a bumblebee (Heinrich, 1975). The above examples may point
out some of the range of possible interactions at the ecosystem level
that should be examined more closely in terms of energetics.

I conclude that there are large possibilities for energetic interrelationships in pollination biology at the ecosystem level. At the
present time, however, most of these possibilities are relatively
unexplored and the study is in its infancy.

References

Baker, H.G., Cruden, R.W., Baker, I.: Minor parasitism in pollination
   biology and its community function: The case of *Ceiba acuminata*.
   BioSci. 21, 1127-1129 (1971)
Bawa, K.S.: Breeding systems of tree species of a lowland tropical
   community. Evolution 28, 85-92 (1974)
Butler, C.G.: The behaviour of bees when foraging. J. Roy. Soc. Arts
   93, 501-511 (1945)
Carpenter, L.: Plant-pollinator interactions in Hawaii: Pollination
   energetics of *Metrosideros collina* (Myrtaceae). Ecology 57g 1125-1144
   (1976)
Frankie, G.W., Baker, H.G.: The importance of pollinator behavior in
   the reproductive biology of tropical trees. An. Inst. Biol. Univ.
   Nal. Auton. Mexico 45 Ser. Botanica (1), 1-10 (1974)
Free, J.B., Spencer-Booth, Y.: The foraging areas of honeybee colonies
   in fruit orchards. J. Hort. Sci. 38, 129-137 (1963)
Gill, F.B., Wolf, L.L.: 1975. Economics of feeding territoriality in
   the Golden-winged sunbird. Ecology 56, 333-345 (1975)
Grant, V., Grant, K.A.: Flower Pollination in the Phlox Family. New
   York: Columbia Univ., 1965
Hainsworth, F.R., Wolf, L.L.: Energetics of nectar extraction in a
   small, high altitude, tropical hummingbird, *Selasphorus flammula*.
   J. Comp. Physiol. 80, 377-387 (1972)
Heinrich, B.: Bee flowers: A hypothesis on flower variety and bloom-
   ing times. Evolution 29, 325-334 (1975)
Heinrich, B.: The foraging specializations of individual bumblebees.
   Ecol. Mon. 46, 105-128 (1976a)
Heinrich, B.: Flowering phenologies: Bog, woodland, and disturbed
   habitats. Ecology 57, 890-899 (1976b)
Heinrich, B.: Resource partitioning among some eusocial insects:
   Bumblebees. Ecology 57, 874-889 (1976c)
Heinrich, B., Raven, P.H.: Energetics and pollination ecology. Science
   176, 597-602 (1972)
Janzen, D.H.: Euglossine bees as long-distance pollinators of tropical
   plants. Science 171, 203-205 (1971)

Judd, W.W.: Studies of the Byron bog in southwestern Ontario. II. The succession and duration of blooming of plants. Can. Field-Naturalist 72, 119-121 (1958)

Knuth, P.: Handbook of Flower Pollination (Transl. J.R.A. Davis). Oxford: Oxford Univ., 1906-1909, 3 Vols., 1729 pp

Levin, D.A.: Low frequency disadvantage in the exploitation of pollinators by corolla variants in Phlox. Am. Naturalist 106, 453-460 (1972b)

Levin, D.A.: Pollen exchange as a function of species proximity in Phlox. Evolution 26, 251-258 (1972a)

Levin, D.A., Berube, D.E.: Phlox and Colias: The efficiency of a pollination system. Evolution 26, 242-250 (1972)

Lewis, H.: Experimental sympatric populations of Clarkia. Am. Naturalist 95, 155-168 (1961)

Macior, L.W.: Behavioral aspects of coadaptations between flowers and insect pollinators. Ann. Miss. Bot. Garden 61, 760-769 (1974)

MacSwain, J.W., Raven, P.H., Thorp, R.W.: Comparative behavior of bees and Onagraceae. IV. Clarkia bees of the western United States. Univ. Cal. Publ. Ent. 70, 1-80 (1973)

Mosquin, T.: Competition for pollinators as a stimulus for the evolution of flowering time. Oikos 22, 398-402 (1971)

Müller, H.: The Fertilization of Flowers (Transl. and ed. D.A.W. Thompson). London: MacMillan, 1883, 669 pp

Pojar, J.: Reproductive dynamics of four plant communities of southwestern British Columbia. Can. J. Botany 52, 1819-1834 (1974)

Reader, R.J.: Competitive relationships of some bog ericads for major insect pollinators. Can. J. Botany 53, 1300-1305 (1975)

Small, E.: Insect pollinators of the Mer Bleue peat bog of Ottawa. Can. Field-Naturalist 90, 22-28 (1976)

Snow, D.W.: A possible selective factor in the evolution of fruiting seasons in tropical forest. Oikos 15, 274-281 (1965)

Snow, D.W.: Evolutionary aspects of fruit-eating by birds. Ibis. 113, 194-202 (1971)

Stiles, F.G.: Ecology, flowering phenology, and hummingbird pollination of some Costa Rican Heliconia species. Ecology 56, 285-301 (1975)

Thien, L.B., Marcks, B.G.: The floral biology of Arethusa bulbosa, Calopogon tuberosus, and Pogonia ophioglossoides. Can. J. Botany 50, 2319-2325 (1972)

Wilson, D.S.: Evolution on the level of communities. Science 192, 1358-1360 (1976)

Wolf, L.L.: The impact of seasonal flowering on the biology of some tropical hummingbirds. Condor 72, 1-14 (1970)

Terrestrial Saprophagous Arthropods

Chapter 6

The Roles of Terrestrial Saprophagous Arthropods in Forest Soils: Current Status of Concepts

D. A. CROSSLEY, JR.

Introduction

Recognition of the importance of soil invertebrates antedates the beginnings of scientific ecology. Darwin's early work established the importance of earthworms. The classic works of Bornebusch (1930) and Jacot (1940) are touchstones which modern researchers on soil fauna still use to place their own studies into a larger perspective. Even as modern ecology developed its large-ecosystem, holistic viewpoints, the importance of soil arthropods has continued to be accepted (O'Neill, 1971; Chew, 1974).

The roles of arthropods in forest soil-litter systems have been envisioned in various ways. In general, arthropods are viewed as regulators of the decomposition segment of forest ecosystems. "Decomposition" here is accepted to be an ecosystem-level process, sensu Odum (1971). From a nutrient flow viewpoint, soil arthropods are envisioned as accelerating (or delaying) nutrient release from decomposing organic matter. They may do this directly: by feeding upon organic matter and associated microflora; or indirectly: by channeling and mixing of the soil, improving quality of substrate for microflora, inoculation of organic debris with microbes, selective grazing upon microflora, preventing senescence of microfloral populations, and so forth. It will be noted that these effects are largely anecdotal because most are difficult to quantify in a satisfactory or meaningful way.

Holism should caution us against viewing soil arthropods solely in terms of the decomposition process. The soil fauna interacts with many other components of ecosystems. For example, the abundant soil predator fauna attacks herbivorous insects which spend a portion of their life history in the soil. Thus, an interrelationship can be shown between decomposition-based food chains and green foliage food chains (Crossley, 1970). Also, forest vertebrates ("wildlife") may be influenced by soil arthropods. The two groups may compete for food (mushrooms), and soil arthropods may serve as intermediate hosts for parasites of vertebrates (Dindal, 1971).

Direct Effects of Soil Arthropods upon Decomposition

The direct effects of saprophagous arthropods—that is, consumption of organic debris and microflora—are still the prominent conceptual ones. Ingestion rates and food web flows are key variables which have been measured in much recent research on soil arthropods (MacLean, 1974; Cornaby et al., 1975; and others). Research now includes such approaches as radioisotope techniques (Kowal and Crossley, 1971; McBrayer and Reichle, 1971), and mathematical compartment modeling of soil arthropod food webs and their substrates (Gist and Crossley, 1975). Generally such techniques focus on precise measurement of the flow of energy or nutrients among compartments.

Pioneering work considered consumption by saprophages in terms of dry matter. Early workers estimated saprophage consumption to be about 20-30% of the annual dry matter input to the forest floor (Ulrich, 1933). Since about half of the litter input disappeared annually, many reasoned that soil animals were responsible for about 50% consumption of the dry matter input.

With the emergence of energy flow as a unifying principle, the role of soil animals in forest decomposition became deemphasized because soil arthropods are less abundant than the soil microflora, and have slower metabolic and turnover rates. The great majority of soil respiration is due to microbial activities. Consequently, some recent studies have estimated that soil fauna contributed less than 1% to the annual CO_2 evolution from forest soils (McBrayer et al., 1974). Since direct consumption appears to be very low, authors have attributed indirect regulation, through microfloral-faunal interactions, as a major role for soil arthropods (Chew, 1974).

Nutrient cycling has emerged as a concept to rival energetics for describing trophic relationships (Todd et al., 1973; Gist and Crossley, 1975). Measurement of nutrient flow in food chains can yield several independent estimates of ingestion rates, compared to one estimate from energetics (Crossley, 1969). Although the results of nutrient flow studies are still preliminary and sometimes contradictory (Kowal and Crossley, 1971), the overview emerging appears to be a reassertion of the importance of direct feeding by soil arthropods. Based on calcium flow in a food web of hardwood forest floor saprovores, Gist and Crossley (1975) estimated that 20% of litter input was consumed. This estimate approaches the Bornebusch (1930) and Ulrich (1933) values. Using potassium as a food web tracer yielded lower estimates of bio-mass consumption (7.5%). Potassium in litter is far more leachable than is calcium, so it is a possible source of discrepancy between the two nutrients. These nutrient flow estimates considered organic debris plus microflora as a single food source. Microflora would be more realistically considered to be an intermediary. Nevertheless, results suggest that direct feeding is a significant role for soil arthropods.

One difficulty with measurements of either energetics or nutrient flow in food chains has been in the assessment of the proportion of injested substance which actually becomes assimilated. In energetics, estimation of the fraction assimilated is difficult and usually is one of the weaker parts of a budget (Crossley and Van Hook, 1970; Luxton, 1972). Early work with radioactive tracers in soil arthropods (Reichle and Crossley, 1965) overestimated assimilation. More refined techniques and analytical models (Reichle, 1969; Goldstein and Elwood, 1971) are now available, and all evidence indicates that assimilation of both nutrients and energy by soil arthropods is low. If so, the discrepancy between estimates of ingestion and CO_2 evolution may become resolved, at least in part. It is noteworthy that different nutrients (and energy) may have very different assimilation efficiencies and so there is no single assimilation fraction (Van Hook and Crossley, 1969). Some authors have suggested that assimilation fractions have been underestimated (Coleman et al., 1976).

Current thinking then, is that direct effects of soil arthropods upon the decomposition process are of considerable importance, but information is not adequate for a definitive assessment. Information now being developed (some reported in this volume) will help to clarify trophic relationships between soil arthropods, microflora, and substrates.

Indirect Effects of Soil Arthropods upon Decomposition

The consensus is that indirect effects of soil arthropods on decomposition and fertility in general may exceed their direct effects of feeding on litter and microflora. Lumped together as indirect effects are such phenomena as conversion of litter to feces, fragmentation of litter, mixing of litter and soil, regulation of microflora, and others. Quantification by observation is difficult, but experimentation demonstrates that such effects are significant.

Kurcheva (1960) provided a dramatic demonstration of the influence of soil arthropods on weight loss by *Quercus robur* leaf litter. After 140 days on plots treated with napthalene to exclude soil fauna, litter retained 91% of its original weight. In contrast, on control plots with fauna, litter retained only 45% of its initial weight.

Witkamp and Crossley (1966) repeated Kurcheva's experiment, using white oak, *Quercus alba*, leaf litter. Radioactive tracer[134]Cs was incorporated into leaves, which were confined in mesh litter bags to permit measurement of weight loss through time. Less dramatic results were obtained after 50 weeks; litter in naphthalene-treated plots retained 55% of its initial weight, while litter in control plots retained 40%. Litter in naphthalene-treated plots retained 15% of tracer[134]Cs, compared with 6% in controls. Analysis showed that litter with or without soil arthropods lost weight at rapid rates during initial stages of the experiment. Divergence between naphthalene treatment and control occurred after the readily soluble fractions had been leached out. The large difference in[134]Cs retention was attributed to a fragmentation effect. Presumably the soil fauna fragmented the litter, exposing a greater surface area to the effects of leaching by rainwater. Earlier, van der Drift (1958) had emphasized the role of soil fauna in mechanical fragmentation of litter.

Lee (1974) studied litter decomposition in forested plots treated with mirex, an insecticide used in the southeastern United States to combat the imported fire ant. Results opposite to those obtained in naphthalene experiments were observed: litter in mirex-treated plots lost weight more rapidly than in control plots. Among soil arthropods, only ant and centipede populations were reduced. Results might be interpreted as a lessening of predator pressure upon the saprophages, but no increase in saprophage populations could be demonstrated. Some effect on the microfloral complex may have been induced by the chemical treatment, but the question remains unresolved.

Nylon mesh bags are another device used for exclusion of soil fauna (Crossley and Hoglund, 1962; Edwards and Heath, 1963). Different mesh sizes excluded different size classes of soil fauna, so that a crude separation of feeding by microfauna, mesofauna and macrofauna was possible. Bags with very small mesh sizes (0.05 mm) presumably allowed only bacteria, fungi and microfauna to invade the litter inside. Virtually no decomposition was reported with such small mesh bags in a West Africa rain forest. In intermediate-sized mesh bags which allowed microorganisms and mesofauna (mites and collembolans) to enter, decomposition proceeded rapidly. Little difference was found between intermediate and coarse-meshed bags in rates of litter disappearance.

These experiments and others (Williams and Wiegert, 1971) demonstrate in a gross way the magnitude of impact by soil fauna, an impact which often exceeds consumption alone. Chew (1974) termed such experiments as crude, and cautioned that they must be interpreted with caution.

However, perturbations of various sorts seem to be a legitimate experimental method for investigating roles of soil arthropods.

Conversion of leaf litter to arthropods feces is a form of substrate
modification. Since assimilation of ingested material is indeed low,
soil arthropods might be viewed as "feces machines," and the generation
of feces might be a major role for arthropods in forest soils. The
importance of arthropod feces in the decomposition process is addressed in Chapter 7 of this volume. In general, feces have been reported to
lose weight more rapidly than parent litter. This finding may not be
universal (Nicholson et al., 1966). Feces are different from parent
litter in several respects: higher in pH, greater capacity to retain
moisture, increased surface-to-volume ratios, and in general, occurring in a form more suitable for colonization by bacteria than fungi
(McBrayer, 1973). The tunnels of passalid beetles have been characterized as an "external rumen." This beetle, which tunnels in decaying
logs on the forest floor, was shown to be dependent upon the reinges-
tion of its own feces. The feces were described as little more than
ground wood, on which a microflora rapidly developed. The analogy is
inaccurate, but it is intriguing to characterize the forest floor as
a giant arthropod rumen.

The dynamics of microfloral and faunal interactions are probably the
crux of the indirect influence of soil arthropods on the decomposition process. This interaction is a difficult subject for experimentation even under laboratory conditions. Feeding habits of fungivorous
arthropods remain poorly known. The literature contains conflicting
reports on the feeding preferences of fungivorous mites in laboratory
cultures. Mitchell and Parkinson (1976) found that field-collected
oribatid mites contained fungal material which the mites would not
eat in laboratory cultures. Given these contradictions, it would appear that factors influencing feeding habits of fungivores is a field
requiring innovative experimentation and observation.

The microcosm experiments of Witkamp and associates have addressed
microfloral-faunal interactions in a systems context (Patten and
Witkamp, 1967). In experiments using radioactive[134]Cs as a tracer, they
followed the microbial immobilization of nutrients. Systems without
millipedes immobilized 24% of [134]Cs from decomposing litter, whereas
systems with millipedes immobilized only 11% of the [134]Cs. In further
microcosm experiments, Ausmus and Witkamp (1973) demonstrated that
microcosms containing mites maintained a higher, more stable microbial
biomass than did control microcosms lacking mites.

Finally, it may be unproductive to attempt to separate direct and
indirect effects of soil arthropods on decomposition processes. Experimental approaches such as the field exclusion work and laboratory
microcosm studies both suggest that the distinction between direct and
indirect effects may not be important. Decomposition results from an
interaction between physical and biological processes, with regulation
accomplished by biota. Short feedback loops for nutrient flow (microbial immobilization, consumption of microbes by fauna) may be sufficient regulatory mechanisms. Figure 1 shows a conceptual model for
nutrient flow, which separates physical and biological processes. Input
is raw litter which must be weathered by physical processes before it
becomes suitable for microbial or faunal attack. The importance of
initial physical weathering varies among deciduous litter species. Once
decomposition is initiated, nutrients may continue to be removed by
physical processes, or by microfloral or faunal feeding. Eventually,
nutrients reach an inorganic state and may be lost from the system.
Several feedback loops occur; for example, nutrients accumulated by

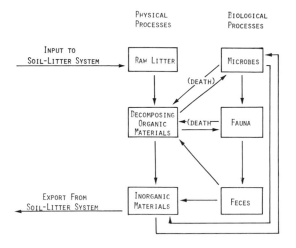

Fig. 1. Compartment model of nutrient transport through soil decomposition system (modified from Crossley, 1976)

microbes may be returned via fauna to the pool of decomposing organic matter. The importance of various pathways doubtless varies with season, and differs for forest types and nutrients considered. The significance of some of the feedback loops is the subject of current experimentation.

Emerging Concepts of Roles

Concepts of the roles of soil arthropods should sharpen as information accumulates. Hopefully, we will reach a better understanding of how soil fauna regulate decomposition processes, when and under what conditions such regulation operates, and how man's manipulations influence regulation. As our concepts sharpen, it is likely that entirely new perspectives will emerge on the roles of soil arthropods in forest ecosystems.

Much current research at ecosystem levels is using nutrient cycling as a unifying theme. There is already considerable evidence to show that soil fauna play important and unique roles in nutrient cycling. Cromack and associates (this volume) are proposing that soil fauna are important in regulating calcium oxalate degradation in soil systems.

The soil arthropod community is a mixture of r- and K-specialists. Shifts in the composition of this mixture must be significant in the regulation of decomposition. However, the question has not been addressed experimentally. It is noteworthy that the soil microarthropod r-specialists (collembolans) and K-specialists (oribatid mites) seemingly compete for the same food resource (fungal hyphae). Collembolan feeding may be prominent in periods of rapid mycelial growth, while oribatids may utilize senescent microflora (possibly with high calcium concentrations).

How do soil arthropod activities influence the rhizosphere? In particular, do arthropods influence the distribution and abundance of mycorrhizae, so important to forest trees (Marx and Bryan, 1975)? It would appear that soil arthropods have the capacity to influence, perhaps to regulate, nutrient uptake by plants.

Some regulation of the decomposition process results from the production of secondary plant substances, such as tannins (Feeny, 1970). Litter containing such products may be slow to decompose until feeding deterrents are broken down or lost via leaching. The abilities of some soil arthropods or their microbial associates to metabolize aromatic compounds may define their particular role in soil systems (Neuhauser et al., 1974).

Forest ecosystems are subject to a variety of manipulations designed to enhance productivity and economic value. The effects of various silvicultural practices upon soil mesofauna has been reviewed recently (Hill et al., 1975). Eventually, knowledge of the roles of soil arthropods may require that silvicultural practices be modified, leading to improved forest management.

Acknowledgments. Research supported by the Eastern Deciduous Forest Biome, US-IBP, funded by the National Science Foundation under Interagency Agreement AG-199, BMS76-00761 with the Energy Research and Development Administration-Oak Ridge National Laboratory.

References

Ausmus, B.S., Witkamp, M.: Litter and soil microbial dynamics in a deciduous forest stand. EDFB-IBP-73-10. Oak Ridge Nat. Lab., 1973, 183 p.

Bornebusch, C.H.: The fauna of the soil. Forstl. Forsgsr. Danm. 11, 1-224 (1930)

Chew, R.M.: Consumers as regulators of ecosystems: an alternative to energetics. Ohio J. Sci. 74, 359-370 (1974)

Coleman, D.C., Hunt, H.W., Leetham, J.W.: Partitioning of saprophytic energy-flow in grassland ecosystems. Proc. 4th Soil Microcommunities Conf. Dindal, D.L. (ed.). 1976, in press.

Cornaby, B.W., Gist, C.S., Crossley, D.A., Jr.: Resource partitioning in leaf-litter faunas from hardwood-converted to pine forests. In: Mineral Cycling in Southeastern Ecosystems, Howell, F.G., Smith, M.H. (eds.). ERDA Symp. Ser. (CONF. 740513), 1975, pp. 588-597

Crossley, D.A., Jr.: Comparative movement of ^{106}Ru, ^{60}Co and ^{137}Cs in arthropod food chains. In: Symp. Radioecology, Nelson, D.J., Evans, F.C. (eds.). Washington: USAEC, 1969, pp. 687-695

Crossley, D.A., Jr.: Roles of microflora and fauna in soil systems. In: Pesticides in the Soil: Ecology, Degradation and Movement. East Lansing: Michigan State Univ., 1970, pp. 30-35

Crossley, D.A., Jr.: Oribatid mites and nutrient cycling. Proc. Symp. Soil Mites. Philadelphia: (1976, in press)

Crossley, D.A., Jr., Hoglund, M.P.: A litter-bag method for the study of microarthropods inhabiting leaf litter. Ecology 43, 571-573 (1962)

Crossley, D.A., Jr., Van Hook, R.I., Jr.: Energy assimilation by the house cricket, *Acheta domesticus*, measured with radioactive Chromium-51. Ann. Ent. Soc. Am. 63, 512-515 (1970)

Dindal, D.L.: Review of soil invertebrate symbiosis. In: Proc. First Soil Microcommunities Conf. Dindal, D.L. (ed.). Washington: USAEC. 1971, pp. 227-256

van der Drift, J.: The role of the soil fauna in the decomposition of forest litter. In: Proc. 15th Intern. Cong. Zool. 1958, pp. 357-360

Edwards, C.A., Heath, G.W.: The role of soil animals in breakdown of leaf material. In: Soil Organisms. J. Doeksen, J., van der Drift, J. (eds.). Amsterdam: North Holland, 1963, pp. 76-84

Feeny, P.: Seasonal changes in oak leaf tannins and nutrients as a
    cause of spring feeding by winter moth caterpillars. Ecology 51,
    565-581 (1970)
Gist, C.S., Crossley, D.A., Jr.: A model of mineral cycling for an
    arthropod foodweb in a Southeastern hardwood forest litter com-
    munity. In: Mineral Cycling in Southeastern Ecosystems. Howell, F.G.,
    Smith, M.H. (eds.). ERDA Symp. Ser. (CONF. 740513), 1975a, pp.84-106
Gist, C.S., Crossley, D.A. Jr.: Feeding rates of some cryptozoa as
    determined by isotopic half-life studies. Environ. Entomol. 4(4),
    625-631 (1975b)
Goldstein, R.A., Elwood J.W.: A two-compartment, thre-parameter model
    for the absorbtion and retention of ingested elements by animals.
    Ecology 52, 935-939 (1971)
Hill, S.B., Metz, L.J., Farrier, M.H.: Soil mesofauna and silvicultural
    practices. In: Forest Soils and Forest Land Management. Bernier, B.,
    Winget, C.H. (eds.). Quebec: Univ. Laval, 1975, pp. 119-135
Van Hook, R.I. Jr., Crossley, D.A., Jr.: Assimilation and biological
    turnover of Cesium-134, Iodine-131, and Chromium-51 in brown crick-
    ets, Acheta domesticus (L.) Health Phys. 16, 463-467 (1969)
Jacot, A.P.: The fauna of the soil. Quart. Rev. Biol. 15, 28-57 (1940)
Kowal, N.E., Crossley, D.A., Jr.: The Ingestion rates of microarthro-
    pods in pine mor, estimated with radioactive calcium. Ecology 52,
    444-452 (1971)
Kurcheva, G.F.: The role of invertebrates in the decomposition of oak
    leaf litter. Pocroredenie 4, 16-23 (1960)
Lee, B.J.: Effects of mirex on litter organisms and leaf decomposition
    in a mixed hardwood forest in Athens, Georgia. J. Environ. Qual.
    3, 305-311 (1974)
Luxton, M.: Studies on the oribatid mites of a Danish beech wood soil.
    I. Nutritional biology. Pedobiologia 12, 434-463 (1972)
MacLean, S.F., Jr.: Primary production, decomposition, and the activity
    of soil invertebrates in tundra ecosystems: A hypothesis. In: Soil
    Organisms and Decomposition in Tundra. Holding, A.J. (ed.).
    Stockholm: 1974, pp. 197-206
Madge, D.S.: Leaf fall and litter disappearance in a tropical forest.
    Pedobiologia 5, 273-288 (1965)
Marx, D.H. Bryan, W.C.: The significance of mycorrhizae to forest
    trees. In: Forest Soils and Forest Land Management. Bernier, B.,
    Winget, C.H. (eds.). Quebec: Univ. Laval, 1975, pp. 107-117
Mason, W.H., Odum, E.P.: The effect of coprophagy on retention and
    bioelimination of radionuclides by detritus-feeding animals. In:
    Proc. 2nd Symp. Radioecology. Nelson, D.J., Evans, F.C. (eds.).
    Washington: USAEC 1969, pp. 721-724
McBrayer, J.F.: Exploitation of deciduous leaf litter by Apheloria
    montana (Diplopoda: Eurydesmidae). Pedobiologia 13, 90-98 (1973)
McBrayer, J.F., Reichle, D.E.: Trophic structure and feeding rates of
    forest soil invertebrate populations. Oikos 22, 381-388 (1971)
McBrayer, J.F., Reichle, D.E., Witkamp, M.: Energy flow and nutrient
    cycling in a cryptozoan food-web. Oak Ridge Natl. Lab. EDFB-IBP-
    73-8, 78 p. (1974)
Mitchell, M.J., Parkinson, D.: Fungal feeding of oribatid mites (Acari:
    Cryptostigmata) in an aspen woodland soil. Ecology 57, 302-312 (1976)
Neuhauser, E., Youmell, C., Hartenstein, R.: Degradation of benzoic
    acid in the terrestrial crustacean, Oniscus asellus. Soil Biol.
    Biochem. 6, 101-107 (1974)
Nicholson, P.B., Bocock, K.L., Heal, O.W.: Studies on the decomposi-
    tion of faecal pellets of a millipede (Glomeris marginata (Villers)).
    J. Ecol. 54, 755-766 (1966)
Odum, E.P.: Fundamentals of Ecology. Philadelphia: W.B. Saunders,
    1971, 574 p
O'Neill, R.V.: Systems approaches to the study of forest floor arthro-
    pods. Sys. Anal. and Stimulation Ecol. 1, 441-477 (1974)

Patten, B.C., Witkamp, M.: Systems analysis of ^{134}Cs kinetics in terrestrial microcosms. Ecology 48, 803-809 (1967)

Reichle, D.E.: Measurement of elemental assimilation by animals from radioisotope retention patterns. Ecology 50, 1102-1104 (1969)

Reichle, D.E., Crossley, D.A., Jr.: Radiocesium dispersion in a cryptozoan food web. Health Phys. 11, 1375-1384 (1965)

Todd, R.L., Cromack, K., Jr., Stormer, J.C., Jr.: 1973. Chemical exploration of the microhabitat by electron microprobe analysis of decomposer organisms. Nature (London) 243, 544-546 (1973)

Ulrich, A.T.: Die macrofauna der Waldstrev. Mitt. Forst., Forstwiss. 4, 283-323 (1933)

Williams, J.E., Wiegert, R.G.: Effects of naphthalene application on a coastal plain broomsedge (Andropogon) community. Pedobiologia 11, 58-65 (1971)

Witkamp, M., Crossley, D.A., Jr.: The role of arthropods and microflora in breakdown of white oak litter. Pedobiologia 6, 293-303 (1966)

Chapter 7

Regulation of Deciduous Forest Litter Decomposition by Soil Arthropod Feces

D. P. WEBB

Introduction

Soil invertebrates may consume 20-100% of annual litter input and in so doing produce an immense amount of excrement (Kurcheva, 1960; Jongerius, 1963; Anderson and Healey, 1970).

Feces do not differ significantly chemically from parent litter (Edwards, 1974). In fact, microarthropod feeding preferences for specific litter components are not altered after litter has passed through macroarthropod guts (Dunger, 1958). Presumably, feces merely represent pulverized litter which offers greater surface area for leaching and microbial attack (Englemann, 1961; Kevan, 1962; Schaller, 1968; Wallwork, 1970; Ausmus and Witkamp, 1973; Jensen, 1974). Positive feedback between microflora and soil fauna (Dunger, 1958) is believed to produce a slow step by step humification of litter as soil, litter, feces and microflora are ingested and reingested (Kurcheva, 1960; Striganova, 1971). Small amounts of energy are extracted and minerals are leached or microbially mobilized, until only humic materials remain. This humus consists primarily of disintegrated fecal pellets (van der Drift, 1951; Harding and Stuttard, 1974). The process is extremely slow in all but nonrecalcitrant litter (low C/N ratio) such as that from *Fagus*, *Castanea*, and *Cornus* (Dunger, 1958). Such litter decomposes rapidly via physical processes and, therefore, exclusion of arthropods from it has failed to show any decrease in decomposition rates (Edwards and Heath, 1963; Harding and Stuttard, 1974).

The success of this step by step process depends on differences between litter, feces, and soil conglomerates in decomposition processes. The purpose of this study was to provide information on the effects of arthropod feces on rates of the following decomposition processes: (1) nitrogen fixation, (2) microfloral production (ATP), (3) microarthropod production, (4) substrate weight loss, and (5) substrate mineral loss.

Methods and Materials

Study Site

Field experiments were conducted on a deciduous Appalachian mixed hardwood watershed (WS 2) of the Coweeta Hydrologic Laboratory (Dils, 1957) near Franklin, North Carolina. Watershed 2 is a south facing control watershed of 16 ha (slope approximately 30°, 730 to 975 m elevation) which has not been disturbed by logging for 50 years. It is bordered by a white pine watershed and an experimental logging watershed.

Experimental Approach

I used four approaches to examine the effect of arthropod feces on decomposition processes:

1. Examination of feces versus litter as microhabitats for microarthropods and microflora

2. Enrichment of plots with feces to determine large scale effects on processes in litter and soil

3. Comparison of fecal substrate results (#1) with enrichment results (#2) to determine any indirect catalytic effects

4. Extension of experimental feces model to other detritivorous arthropod feces.

Feces Factory

I chose feces of the millipede, *Narceus annularis* (Spirobolidae), as the model macroarthropod feces because of their pellet-like form, high rate of production (Shaw, 1970) and the fact that millipedes are the dominant litter consuming arthropods in North American deciduous forests. To obtain feces for experimentation, I constructed a "feces factory:" six 6-in deep, 2 ft x 4 ft wooden boxes were lines with plastic, filled with mixed old litter (L2 layer) and 150 millipedes and stacked one upon another. Boxes were changed once a week to minimize decomposition and ingestion of feces. The litter-feces mixture was sifted twice over 6.2 mm hardware cloth, yielding approximately 80% feces and 20% orts (scraps). This mixture was stored at 4°C until used. In this manner I collected approximately 18.5 kg dry weight (233% moisture content) of feces (approximately 4 million pellets at 5mg/pellet-3.5 x 4.5 mm average dimensions).

Field Substrate Experiments (Experimental Approach 1)

1 dm^2 fiberglass mesh bags (1 mm^2 openings) containing 5-10 g dry weight of substrate were inserted into the F layer of a row of six plots. Substrates included (1) *Narceus* feces (particle size >3.1 mm) and (2) mixed old litter, or ground old litter (three particle sizes: <.5 mm, .5-1 mm, 1-2 mm) which was rewet after weighing and compressed into conglomerates (8 cm diameter by .6 cm thick). Substrates (two bags from each of three plots per substrate) were monitored biomonthly for moisture content, nitrogen fixation (Hardy et al., 1968), ATP, microarthropod numbers, weight loss and mineral loss.

Field Enrichment Experiments (Experimental Approach 2)

In order to estimate gross effects of feces on decomposition processes, I enriched three sets of square plots (2 x 2 m) with 690 dry g/m^2, 345 dry g/m^2 or no *Narceus* feces in October (before leaf fall) of 1974. The enrichment mimics a summer crop of feces.

Two weeks later, I collected newly senescent *Quercus alba* foliage and placed 5 g dry wt in 1 dm^2 fiberglass mesh bags which were placed in a medium position within the newly fallen L layer on the 12 plots. I monitored moisture content, microarthropod numbers, weight and mineral loss in two bags per plot at monthly intervals. In addition I also monitored two soil cores per plot (4.5 cm diameter by 5 cm depth) at the same intervals for moisture content, ATP, microarthropod numbers, and nitrogen fixation.

Laboratory Fecal Leachate Experiment

Water which percolates through litter or feces carries off microbes and chemicals to soil beneath the leached substrate. This process was approximated by enriching ground old litter (1-2 mm) with leachate from (1) *Narceus* feces and/or (2) old litter.

I prepared leachates by immersing the appropriate substrates in deionized water for 24 h (20 ml of water to each dry gram substrate). Cylindrical plastic containers were each filled with 2 dry g of ground old litter to which leachate was added. Eight replicates of litter leachate (8 ml), feces leachate (8 ml) and litter (4 ml) + feces (4 ml) leachate yielded 24 total cultures. These were incubated at room temperature in darkness within a plastic shoebox containing moist sponges. ATP content was estimated at 1, 5, and 29 days.

Results and Discussion

Field Substrate Experiments

Ash content of feces (16%) was greater than litter (9%) initially, but this difference decreased throughout the year due to greater C mineralization (organic matter loss) in litter. Fe, Mg, B, Zn, Al, Ba, and Ca were initially more concentrated in feces, but loss and gain rates were the same in both substrates (Webb, 1977). Initial fecal and litter concentrations of N, K, Na, P and Cu were the same. Most mineral concentrations showed no overall trend towards loss or gain in one year. Only K showed a significant reduction in concentration (due to initial leaching). P increased in concentration in feces and decreased in litter.

Nicholson et al. (1966) found that *Glomeris* feces increased in ash content in the field. The slightly lower C/N ratio of feces compared with parent litter (Dunger, 1958; Bocock, 1963a) is not enough to account for the consistently higher ash content of feces.

Apparently *Narceus* selectively consumed litter species or components with higher ash content (Lyford, 1943; Edwards, 1974) and passed over items of low ash content such as leaf midribs and large veins (Bocock, 1963a; Shaw, 1970). Although the ash content of feces is higher than litter, the leaching rates of the two substrates are not different.

Both feces and litter had equal nitrogen fixation (Fig. 1) at the time of the first October samples, perhaps due to conditions of low moisture (Fig. 2). In the next two samples, feces fixed more nitrogen than litter.

High nitrogen fixation may explain the reduced C/N ratio and increased NH_3 concentrations (Bocock, 1963a, b; Lodha, 1974) measured in many soil arthropod feces. Bacterial populations do not increase on food passing through millipede digestive tracts (Nicholson et al., 1966) as is common in oliogochaetes (Satchell, 1967; Lofty, 1974). The peak and decline in bacterial activity which occurs during the first two weeks of fecal incubation (van der Drift and Witkamp, 1959; Nicholson et al., 1966; Lodha, 1974) may be due to nitrogen fixing bacteria which quickly utilize initial nutrients or anaerobic conditions in the pellets.

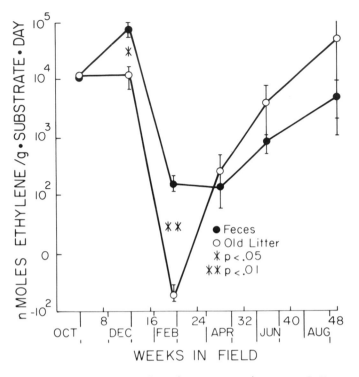

Fig. 1. Acetylene-ethylene reduction assay (mean + S.E., n= 6) for nitrogen fixation performed on field-collected mesh bag enclosed litter and *Narceus* feces. The initially greater fixation in feces may correspond to an initially bacterial peak found in other soil microarthropod feces (Some standard errors are smaller than the radius of the circular symbol and do not show up as bars.)

The initially high ph of feces (Bocock, 1963a; McBrayer, 1973) may also be ephemeral permitting nitrogen fixers to grow for only a limited time, after which feces and litter fix nitrogen at similar rates (Fig. 1).

Despite higher moisture content (Fig. 2), mineral content, and initial nitrogen fixation (Fig. 1); feces had less microflora (ATP) and microarthropod numbers per dry weight. For example, an average of 2.5 times as many microarthropods and 3.0 times as much ATP occurred on litter as on feces. These differences can be explained by differences in surface area to mass ratios of litter and fecal pellets. For example, the surface area: mass ratios of 200 pellets and 125 leaf discs were 130.5 + 3.4 and 358.0 + 7.7 cm^2/dry gram, respectively. Therefore, litter had approximately 2.8 times more surface area than feces.

The pelletized nature of arthropod feces results in a lower surface to mass ratio in macroarthropod feces and higher ratio in microarthropod feces compared with "parent" food litter. Fecal pellets with diameters less than litter lamellar thickness have higher surface area to mass ratios than litter and increase decomposer organism activity by providing more usable substrate. All microarthropods (less than 1 cm long; McBrayer et al., 1974) produce "microfeces" (Harding and Stuttard,

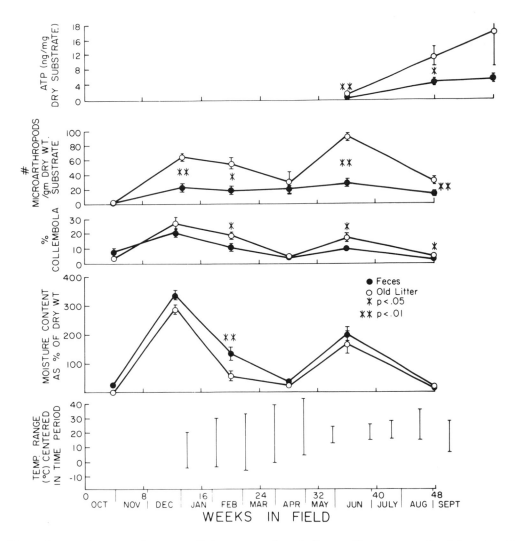

Fig. 2. Estimates (mean + S.E., n = 6) of feces *(Narceus annularis)* and old litter parameters from field collected mesh-bag enclosed substrates. Although feces contained more moisture, the greater surface area : mass ratio of litter allowed a larger standing crop of decomposer organisms per dry weight (Some standard errors are smaller than the radius of the circular symbol and do not show up as bars.)

1974). Fecal pellets with diameters greater than litter lamellar thickness have lower surface area to mass ratios than litter and decrease decomposer organism activity by providing less usable substrate. Most macroarthropod feces would fit into the "macrofeces" category, since macroarthropods are greater than 1 cm long (McBrayer et al., 1974).

Bacterial counts are higher on feces than on parent litter when feces have surface area to volume ratios (SA:V) greater than that of *Narceus* feces (1.7:1). *Glomeris* pellets (3.2:1; calculated from Nicholson et al., 1966) harbored 30 times more bacteria than did litter, and *Eniocyla pul-*

silla (terrestrial trichopteran larva) pellets (6:1, calculated from dimensions in van der Drift and Witkamp, 1959) harbored 1000 times more bacteria than litter (van der Drift and Witkamp, 1959). Assuming an average litter lamellar thickness of .2 mm in litter (SA:V = 10:1), both *Glomeris* and *Eniocyla* pellets should harbor fewer microflora than litter. However, these counts were made during the initial two week bacterial peak, and results may be different in extended field experiments (as in ATP, Fig. 2, vs Fig. 1). Many have assumed that all arthropod pulverization of litter (feces) results in an increase in surface area over the original litter. However, pelletization overrides the effects of pulverization and controls surface area/volume phenomena. In 1957, Nef (Harding and Stuttard, 1974) recognized the importance of pelletization in Phthiracaroid mite fecal pellets. There was a 10,000 fold increase in surface area if constituent (microstructural) particles were considered, but only a four-fold increase due to pelletization.

Romell (1935) worked with the largest macrofeces; i.e., polydesmid excrement clumps. These millipedes excrete a diarheic liquid in a series of droplets which stick together forming conglomerates up to 2 cm in diameter (SA:V = 0.3:1). Romell found less live bacterial cells on this rich organic substrate than in surrounding soil. Although macroarthropods do increase litter disappearance (Edwards and Heath, 1963), disappearance does not necessarily mean that the lost material decomposes more rapidly than the original litter (Crossley, 1970).

Microarthropods have been observed on macrofeces (Harding and Stuttard, 1974), but there are no quantitative studies on this subject. Macrofeces may have fewer microarthropods than "parent" litter due to lower surface area, higher pH (Butcher et al., 1971), lower organic matter (Chernova et al., 1971), higher moisture content (Crossley and Hoglund, 1962), and less microflora (Fig. 2). Predation may also reduce microarthropod numbers because predatory mites are able to search better on substrates having three versus two dimensional geometries (Martin, 1969). Since Collembola are favorite prey of mesostigmatid mites (Karg, 1971), they may be selected against in fecal pellet microhabitats. The data, in fact, show fewer microarthropods and a smaller percentage of Collembola on feces than on parent litter (Fig. 2). Since there is a higher concentration of Ca in feces it may result in a higher percentage of oribatid mites on feces than on litter. The data suggest that both high moisture content and low surface availability reduce the suitability of *Narceus* feces as microarthropod substrates.

Both ATP and microarthropods respond to the available surface area of artificial conglomerates, despite differences in moisture content and constituent particle size (Fig. 3). The graph indicates that a decrease in constituent particle size increases available surface area up to a limit. This limit occurs when smaller particles clump or form persistent conglomerates. Conglomerates formed by larger particles (CG and MG) have little structural integrity because of lower interparticle surface area for cohesion. Tiny ground (TG) conglomerates had greatest structural integrity (the disc appeared whole with cracks as in dried mud, the only obvious disintegration of structural integrity). Although this conglomerate had greatest potential surface area, it had lowest actual available surface area due to high macrostructural integrity induced by its microstructure.

Figure 4 indicates that *Narceus* feces and old litter (FL) lost weight at the same rate. Nicholson et al. (1966) found that *Glomeris* feces and hazel litter lost weight at the same rate (about 50% in one year). *N. annularis* pellets and *Glomeris* pellets have SA:V ratios of approximately 1.7:1 and 3.2:1, respectively. Both constituent particle size and pellet size of feces are in direct proportion to size of the feces producer (Dunger, 1963; Harding and Stuttard, 1974). Although *Glomeris*

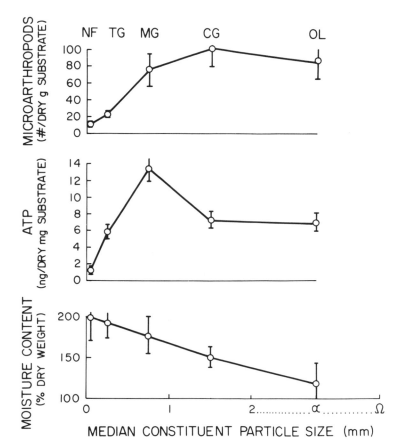

Fig. 3. Estimates (mean + S.E., n = 6) of parameters in mesh bag enclosed substrates (*NF: Narceus* feces; *TG:* tiny ground-, *MG:* medium ground-,
CG: coarse ground artificial conglomerates; *OL:* old litter) collected
in January after four months in field. Although moisture content increases with decreasing constituent particle size, decomposer standing
crop reaches a peak limit and then decreases due to greater interparticle cohesion of smaller constituent particles resulting in stable conglomerate macrostructures with less available surface area (α to Ω is
the unmeasured range of old litter fragment sizes)

fecal pellets have a high available surface area, the tiny constituent
particles produce great interparticle cohesion adding to the pellet's
persistence. On the other hand, *Narceus* fecal pellets have low available surface area but the larger constituent particles have weak interparticle cohesion.

Theoretical Implications

These results suggest that there is a limit to particle size reduction
in soil-litter systems. Macrostructure and microstructure of soil conglomerates are two opposing characteristics controlling decomposition.
Smaller conglomerate size (macrostructure) increases surface area availability, whereas smaller constituent particle size (microstructure)

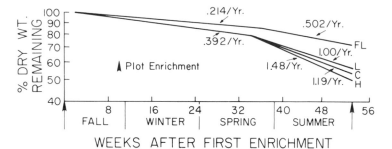

Fig. 4. Weight loss (mean, *n:* 6) of field collected mesh bag enclosed *Narceus* feces and old litter (FL) is the same. Weight loss (mean, *n:* 6) of new *Quercus alba* litter on feces enriched plots *(H:* heavy; *L:* light; *C:* controls) are not significantly different. Second year substrates *(FL)* are more recalcitrant than first year litter

increases interparticle bonding and persistence of macrostructure. Reduction in particle size by faunal feeding, microbial disintegration, and weathering is counteracted by greater interparticle cohesiveness. The smaller the individual particle; the greater the interparticle surface area for water film bonding and adsorption to coagulate the small particles into larger conglomerates. An analogous chemical process has been termed refloculation in aquatic systems (Lust and Hynes, 1973) and will be called reaggregation here.

By processing their feces into compact pellets macro- and microarthropods produce two distinctly different kinds of feces. The macropellets contain constituent particles which are larger than those in micropellets, so the interparticle cohesive forces are weaker. Macrofeces have less available surface area than the original litter and maintain this alteration as long as the litter itself maintains its structural integrity (Fig. 4). Therefore, the primary function of macrofeces is to reduce available substrate surface area.

Micropellets have high available surface area and thus would be expected to decompose rapidly. However, they are composed of extremely tiny particles that have interparticle cohesive forces stronger than in macropellets. This insures their persistence. Microarthropod feces are known to retain their structural integrity even after passage through earthworm guts (Harding and Stuttard, 1974). Oribatid feces have been found in Carboniferous coal seams and Collembolan feces accumulate in layers up to 60 cm in depth in the Alps (Kevan, 1962). However, the tendency of micropellets to filter downward or to be deposited into lower soil profiles does increase their decomposition rates (Jongerius, 1963). The structural persistence of microfeces allows a smaller particle size to exist than would be possible if particles of feces were not forced into a pelletized condition. Therefore, microfeces function to maintain sites where more decomposition per gram of substrate can take place, than if particles were free to reaggregate into large conglomerates in the humic fraction of the soil. I propose that the balance of large to small arthropod activity in soil systems constitutes a regulatory mechanism controlling rates of total decomposition of litter.

Figure 5 summarizes these hypotheses in graphic form. As constituent particle size decreases, decomposition processes (weight loss, mineral loss, microbial and microarthropod activities, etc.) increase. At a

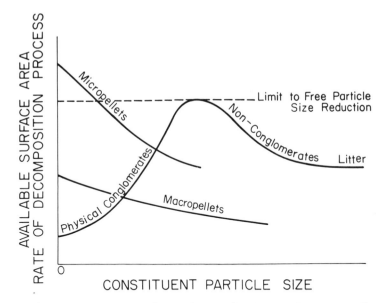

Fig. 5. Graphical representation of physical conglomerate-feces differentiation theory (Webb, 1977). As particle size of litter (*right* to *left*)
is reduced, surface area and decomposition increase until constituent
particles are small enough to aggregate into more stable conglomerates
(limit to free particle size reduction). Physical conglomerates increase
in size as constituent particle size decreases, but arthropod pellets
decrease in size due to the direct relationship of body size to degree
of pulverization and pellet (conglomerate) size. Micropellets are therefore able to maintain a much smaller conglomerate size and break the
limit to free particle size reduction

certain point (limit to particle size reduction) the non-fecal particles
reaggregate and produce free (exposed) particles of a larger size (physical conglomerates). Because micropellets become smaller as constituent
particle size decreases (Dunger, 1963), the resulting conglomerates are
smaller than their physically produced counterparts with the same constituent particle size. This function of microarthropods breaks the
physical limit to free particle size reduction. Just as living organisms work in opposition to the second law of thermodynamics in pumping
entropy out of their systems, soil fauna pump entropy out of conglomerate systems by means of fecal production and, therefore, increase the
diversity and regulate the velocity of surface area phenomena.

Field Enrichment Experiments

Although enrichment of field plots with *Narceus* feces did increase number
of microarthropods/m^2 (Fig. 6); ATP, N-fixation, and moisture content
did not change. The increase in microarthropod densities can be accounted for in all but the first point by microarthropod colonization of
feces that were applied to the enrichment plots. Feces support about
20 microarthropods/g. The significant initial difference cannot be
explained by added substrate (Fig. 2 shows no difference in F and L for
that date). For other dates, subtracting substrate values results in
no synergistic enrichment effect. Therefore, increased microarthropod
as a result of feces enrichment were apparently additive.

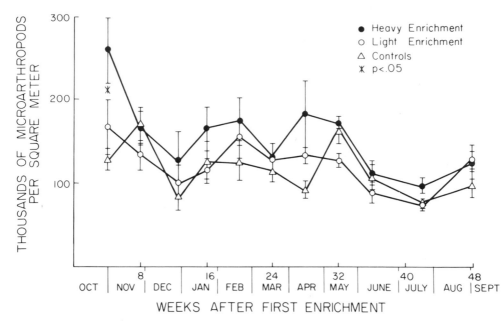

Fig. 6. Numbers of microarthropods (mean + S.E., n: 6) in whole soil-litter systems (soil cores; 5 cm depth, 4.5 cm diameter) from feces enriched plots. Overall differences are due to additional substrate (feces) on the plots. No catalytic effects were monitored here or in any other parametric estimates (ATP, N-fixation, weight loss (Fig. 4), moisture content, fecal leachate experiments)

Ghilarov (1963) has proposed that microflora are enhanced by comminution (earthworm feces). Ghilarov's conclusions were based on comparisons of bacterial counts of earthworm feces (litter incorporated into soil) with those of surrounding soil. Satchell (1967) recognized that these differences could be accounted for entirely on the basis of extra substrate (litter) in the feces.

Enrichment of plots with feces did not change weight loss (Fig. 4, L, C, and H), mineral loss, nor number of microarthropods in new Q. alba litter. Effects of feces on litter decomposition must be seen at the transformed substrate level. Since disintegration of litter and feces were similar (Fig. 4), and assuming that the feces do not catastrophically disintegrate, quantity of particulate input from feces and litter into the lower fermentation layers may be similar. Since feces are higher in ash content due to selective feeding, mineral input is higher from feces at the ecosystem level.

One weakness in this discussion lies in the lack of experimental data on vertical movement of pellets into deeper layers (Witkamp, 1971). Observation on the enriched plots indicated very little downward movement, however. Downward movement of micropellets causes higher disintegration rates and aggregation of micropellets. This would diminish the effectiveness of micropellets as sites of high surface area availability.

Laboratory Fecal Leachate Experiment

Microcosms enriched with litter leachate exhibited initially greater
ATP (at one day) than microcosms enriched with fecal leachate. However after one day there was no difference. Greater initial leaching
of nutrients from the larger surface area in litter could account for
this result. A bacterial peak probably exhausts the leachate nutrients.

Some Unanswered Questions

Transformation of litter into macrofeces reduces the surface area to
mass ratio and therefore removes layers of insulating litter. But the
increase in moisture holding capacity in feces may work to maintain
the insulation of the soil.

Mineral soil is usually incorporated into macroarthropod feces (Dunger,
1958). However, the feces produced in "feces factories" did not have
mineral soil incorporated in them. It remains unknown then if soil
incorporation would have changed loss rates and decomposer activity
per unit of feces area. Mineral soil incorporation could change internal
pellet structure and binding.

Narceus feces lost less than 30% of their weight in one year (Fig. 4).
Loss of weight gradually transforms macropellets into pseudo-micropel-
lets (larger constituent particle size) which may catastrophically
disintegrate sending a pulse input of fine particles into the deeper
soil. Extension of all experiments for a longer period of time could
answer this question.

Acknowledgments. This research was supported by the Eastern Deciduous
Forest Biome, US-IBP, funded by the National Science Foundation under
Interagency Agreement AG-199, BMS760-00761 with the Energy Research
and Development Administration, Oak Ridge National Laboratory. It was
part of the Role of Consumers project administered by Dr. James R.
Kitchell, University of Wisconsin. Work at Coweeta Hydrologic Labora-
tory was performed under a cooperative agreement between the University
of Georgia and the U.S. Forest Service. Partial support for work at
Coweeta was provided by NSF grant GB-43255 to the University of Georgia
(D.A. Crossley, Jr.).

Dr. D. A. Crossley, Jr. provided continual encouragement and advice on
all aspects of the work. My lovely technician, Dell Webb, was of immense
help. Dr. Robert L. Todd (Univ. Georgia) provided assistance with the
microbial analysis.

References

Anderson, J.M., Healey, I.N.: Improvements in the gelatine-embedding
    technique for woodland soil and litter samples. Pedobiologia 10,
    108-20 (1970)
Ausmus, B.S.: The use of ATP assay in terrestrial decomposition studies.
    In: Modern Methods in the Study of Microbial Ecology. Rosswall, T.
    (ed.). Stockholm: Nat. Sci. Res. Council, 1973, pp. 223-34
Ausmus, B.S., Witkamp, M.: Litter and soil microbial dynamics in a
    deciduous forest stand. ORNL Publication EDFB-IBP 73-10, 1973, 183 p.
Bocock, K.L.: The digestion and assimilation of food by *Glomeris*. In:
    Soil Organisms. Doeksen, J., van der Drift, J. (eds.). Amsterdam:
    North-Holland, 1963a, pp. 137-48

Bocock, K.L.: Changes in the amount of nitrogen in decomposing leaf litter of sessile oak. J. Ecol. 51, 555-66 (1963b)

Butcher, J.W., Snider, R., Snider R.J.: Bioecology of edaphic Collembola and Acarina. Ann. Rev. Ent. 16, 249-88. (1971)

Chernova, N.M., Byzova, J.B., Chernova, A.I.: Relationship of number, biomass and gaseous exchange rate indices in microarthropods in substrates with various organic matter contents. Pedobiologia 11, 306-14 (1971)

Crossley, D.A., Jr.: Roles of microflora and fauna in soil systems. In: Pesticides in the Soil: Ecology, Degradation and Movement. Michigan State Univ. 1970, pp. 30-35

Crossley, D.A., Jr., Hoglund, M.P.: A litter-bag method for the study of microarthropods inhabiting leaf litter. Ecology 43, 571-573 (1962)

Dils, R.E.: 1957. A Guide to the Coweeta Hydrologic Laboratory. USDA Forest Service, SE Forest Exp. Sta., Asheville, N.C. Sept. 1957. 40 pp.

van der Drift, J.: Analysis of the animal community in a beech forest floor. Wagenengen: Ponsen and Locifen, 1951, pp. 1-168

van Der Drift, J., Witkamp, M.: The significance of the breakdown of oak litter by *Eniocyla pulsilla*. Burm. Archo. Neerl. Zool. 13, 486-492 (1959)

Dunger, W.: Uber die Veranderung des Fallaubes im Darm von Bodentieien. J. Planz. Boden 82, 174-93 (1958)

Dunger, W.: Leistungsspezifitat bei Streuzersetzern. In: Soil Organisms. Doeksen, J., van der Drift, J. (eds.). Amsterdam: North-Holland, 1963, pp. 92-102

Edwards, C.A.: Macroarthropods. In: Biology of Plant Litter Decomposition. Dickinson, C.H., Pugh, G.J.E. (eds.) London: Academic, 1974, Vol. II, pp. 533-53

Edwards, C.A., Heath, G.W.: The role of soil animals in breakdown of leaf material. In: Soil Organisms, Doeksen, J., van der Drift, J. (eds.). Amsterdam: North-Holland, 1963, pp. 76-84

Edwards, C.A., Reichle, D.E., Crossley, D.A., Jr.: The role of soil invertebrates in turnover of organic matter and nutrients. In: Analysis of Temperate Forest Ecosystems. Reichle, D.E. (ed.). Berlin-Heidelberg-New York: Springer, 1970, pp. 147-172

Englemann, M.D.: The role of soil arthropods in the energetics of an old field community. Ecol. Mon. 31, 221-38 (1961)

Ghilarov, M.S.: On the interrelations between soil dwelling invertebrates and soil microorganisms. In: Soil Organisms. Doeksen, J., van der Drift, J. (eds.). Amsterdam: North-Holland, 1963, pp. 255-259

Harding, J.L., Stuttard, R.A.: Microarthropods. In: Biology of Plant Litter Decomposition. Dickinson, C.H., Pugh, G.J.E. (eds.). London: Academic, 1974, Vol. II, pp. 489-532

Hardy, R.W.F., Holsten, R.D., Jackson, E.K., Burns, R.C.: The acetylene-ethylene assay for N_2 fixation: laboratory and field evaluation. Plant Physiol. 43, 1185-1207 (1968)

Heath, G.W., Edwards, C.A., Arnold, M.K.: Some methods for assessing the activity of soil animals in the breakdown of leaves. Pedobiologia 4, 80-7 (1964)

Jensen, V.: Decomposition of angiosperm tree leaf litter. In: Biology of Plant Litter Decomposition. Dickinson, C.H., Pugh, G.J.E. (eds.). London: Academic, 1974, Vol I, pp. 69-104

Jongerius, A.: Optic-volumetric measurements on some humus forms. In: Soil Organisms. Doeksen, J., van der Drift, J. (eds.). Amsterdam: North-Holland, 1963, pp. 137-48

Karg, W.: 1971. Die freilebenden Gamasina (Gamasides), Raubmilben. VEB Fischer, fena. 1971, 475 pp

Kevan, D.K. McE.: Soil Animals. London: Witherby, 1962, 244 pp

Kurcheva, G.F.: The role of invertebrates in the decomposition of the oak leaf litter. Pocvovedenie 4, 16-20 (1960)

Lodha, B.C.: Decomposition of digested litter. In: Biology of Plant Litter Decomposition, Dickinson, C.H., Pugh, G.J.E. (eds.). London: Academic, 1974, Vol. I, pp. 213-241

Lofty, J.R.: Oligochaetes. In: Biology of Plant Litter Decomposition. Dickinson, C.H., Pugh, G.J.E. (eds.). London: Academic, 1974, Vol. II, pp. 467-488

Lust, D.L., Hynes, H.B.N.: The formation of particles in freshwater leachates of dead leaves. Limn. Oceanogr. 18, 968-77 (1973)

Lyford, W.H.: The palatability of freshly fallen forest tree leaves to millipedes. Ecology 24, 252-61 (1943)

Madge, D.S.: Litter disappearance in forest and savannah. Pedobiologia 9, 288-99 (1969)

Martin, F.J.: Searching success of predators in artificial leaf litter. Am. Midl. Naturalist 81, 218-27 (1969)

McBrayer, J.F.: Exploitation of deciduous leaf litter by *Apheloria montana* (Diplopoda: Eurydesmidae). Pedobiologia 13, 90-98 (1973)

McBrayer, J.F., Reichle, D.E., Witkamp, M.: Energy flow and nutrient cycling in a cryptozoan food-web. Eastern Deciduous Forest Biome, IBP-73-8. 1974

Nicholson, P.B., Bocock, K.L., Heal, O.W.: Studies on the decomposition of the faecal pellets of a millipede (*Glomeris marginata* Villers). J. Ecol. 54, 755-766 (1966)

Raw, F.: Arthropoda (except Acari and Collembola). In: Soil Biology. Burgess, A., Raw, F. (eds.). New York: Academic, 1967, pp. 323-363

Romell, L.G.: An example of myriapods as mull formers. Ecology 16, 67-71 (1935)

Satchell, J.E.: Lumbricidae. In: Soil Biology. Burgess, A., Raw, F. (eds.). New York: Academic, 1967, pp. 259-322

Schaller, F.: Soil Animals. Ann Arbor Sci. Library: Univ. Michigan Press, 1968

Shaw, G.G.: Energy budget of the adult millipede *Narceus annularis*. Pedobiologia 10, 389-400 (1970)

Striganova, B.R.: A comparative account of the activity of different groups of soil invertebrates in the decomposition of forest litter. Ekologiya 4, 36-43 (1971)

Wallwork, J.A.: Ecology of Soil Animals. New York: Mc Graw-Hill, 1970, 283 pp

Webb, D.P.: Roles of soil arthropod feces in deciduous forest litter decomposition processes. Ph.D. dissertation. Univ. Georgia, 1977 166 p.

Witkamp, M.: Soils as components of ecosystems. Ann. Rev. Ecol. Syst. 2, 85-110 (1971)

Chapter 8

Contributions of Cryptozoa to Forest Nutrient Cycles

J. F. McBrayer

Introduction

Satchell (1974), among others, has proposed that decomposer invertebrates may be important in regulating rates of elemental release in
ecosystems where high precipitation, seasonal plant uptake, and low
cation exchange capacity of soils can lead to serious leaching losses.
Satchell indicated that the hypothesis would be difficult to prove and,
in fact, it would be difficult to prove that the animals were not actually competing with vegetation for available ions. Nevertheless, decomposer organisms do exhibit significant elemental concentrations
either above or below that of the plant litter substrate (Reichle,
1971). Further, abundant cations tend to be concentrated little, if
at all, while critical elements which occur in anionic form in the
soil solution, and are most susceptible to leaching losses, tend to
be concentrated most. The following work examines the fluxes of seven
elements, three of which are anionic in the soil solution, through a
decomposer invertebrate food-web.

Methods

Field data were collected from an 0.13 ha *Liriodendron* dominated, cove-
hardwood forest on the U.S. Energy Research and Development Administration Reservation near Oak Ridge, Tennessee. The forest lies within
a sinkhole and is completely surrounded by an oak-hickory forest which
contributes both additional organic matter and water. The added organic
matter from the side slopes, along with the internally produced litter,
supports a microarthropod population density which is ~50% greater than
that of nearby, larger sinkholes which have less external input of
litter.

The invertebrate mesofauna was sampled monthly for one year by means
of 20, 5 cm diameter x 15 cm deep cores taken through the litter layer
and into the mineral soil. The invertebrate macrofauna was sampled
concurrently by means of 10, 0.1 m^2 area circular frames which cut
through the litter to the surface of the mineral soil. Invertebrates
were extracted from the samples using refrigerated Tullgren funnels of
appropriate sizes and were collected in an aqueous solution of 70%
ethanol and 5% glycerin. Densities were calculated from direct counts
of samples and biomass was determined by weighing individuals or groups
of individuals which had been dried to equilibrium with 40°C. Invertebrates to be used for elemental analyses were sorted into predators,
fungivores, and saprovores, washed several times with distilled water,
and dried at 40°C.

Litter standing crops were monitored monthly by means of the samples
taken for macrofauna plus additional samples which were separated into
01, 02, and stems, dried at 40°C, and used in elemental analysis.

Analyses for sodium, magnesium, calcium, potassium, and phosphorous were by atomic absorption spectrophotometry. Nitrogen and sulfur were analyzed by gas chromatography. Litter was analyed six times during the year and invertebrates only once because samples were pooled from all twelve sample dates.

Productivities were calculated daily using the equations of Petrusewicz and Macfadyen (1970) and were based on biomass and field temperatures at each monthly sampling. Daily values were calculated by linear interpolation. All fluxes were calculated as a constant fraction of the compartment caloric equivalent at $10^{\circ}C$ and were scaled to daily field temperatures using a $Q_{10} = 2$. The production equation for saprovores was based on McBrayer (1973):

$$P = C_1 - FU_1 + C_2 - FU_2 - R \tag{1}$$

C_1 represents consumption of O1 litter for which the assimilation efficiency is quite low, 5.2%. O1 litter alone is inadequate to support millipedes. At $10^{\circ}C$, consumption rate was calculated to be .0316 times body caloric equivalence per day. C_2 represents consumption of partially decomposed litter (O2) which has been enriched with microflora and is, consequently, more easily digested and assimilated by the saprovores. Consumption at $10^{\circ}C$ was calculated to be .0299 body caloric equivalence with an assimilation efficiency of 13.8%. Respiration (R) was calculated as .0045 body caloric equivalence per day. Continuous solution of equation 1 would lead to continuous increases in standing crop (ΔB). Since daily standing crop was known, or easily calculated, it was possible to calculate losses (E):

$$P = \Delta B + E \tag{2}$$

Annual net production is calculated by summing E for the year if standing crops return to initial conditions.

For fungivores, only one ingestion term appeared in the production equation:

$$P = C - FU - R \tag{3}$$

Ingestion rate was set at 11.5% of body caloric equivalence per day based on the work of McBrayer and Reichle (1971), corrected for the assimilation efficiency for cesium which they ignored. Assimilation efficiency was set at 60% (Luxton, 1972) and respiration was set at 5.5% per day, based on data compiled from the literature for all taxa (McBrayer et al., 1974). Losses were calculated according to Equation (2).

Predator production and losses were calculated by Equations (3) and (2), respectively. Ingestion rate, 5% per day, and assimilation efficiency, 90%, were based on Van Hook (1971). Respiration rate, 2.85% was based on McBrayer et al. (1974).

Results

Monthly Standing Crops

Invertebrate population minima occurred in March and June with maxima in October and November (Table 1). All three groups followed basically the same pattern with fungivores contributing an average of 45% (40-53% range), predators 38% (30-46% range), and saprovores 17% (12-21% range) of total invertebrate biomass. Total litter standing crop peaked in December but the annual minimum occurred in April, not September. The apparent turnover of 650 $g/m^2/yr$ (December to April) exceeds the autumn litterfall (September to December) and is, itself, an under-estimate of litter disappearance. Litter obviously decomposes most rapidly in summer when there was actually a net accumulation of 450 g/m^2.

Table 1. Monthly standing crops (g/m^2) of forest floor litter and decomposer invertebrates in a *Liriodendron* forest at Oak Ridge, Tennessee, U.S.A.

Month	01 Leaves	01 Stems	02	Saprovores	Fungivores	Predators
January	328	131	220	1.85	3.90	3.12
February	188	74	409	1.38	3.52	2.16
March	214	82	253	1.16	2.41	1.93
April	205	82	123	1.54	4.17	3.81
May	284	99	197	1.13	3.63	3.30
June	283	83	224	0.64	2.20	2.34
July	409	96	365	1.40	3.92	4.58
August	277	149	284	1.38	3.07	3.15
September	267	267	327	1.39	3.48	3.24
October	331	221	368	1.73	5.52	3.97
November	392	186	402	1.99	5.95	3.36
December	466	201	392	1.69	5.17	3.36

Elemental Concentrations and Energy Fluxes

All invertebrates were pooled for analysis of nitrogen and sulfur, resulting in the lack of differences between groups (Table 2). Saprovores concentrated all elements except potassium above substrate levels according to Na >P >N >Ca >S = Mg. Fungivores concentrated elements according to Na >P >K >N >Mg >Ca = S, the latter two having concentration factors approaching one. Predators were equivalent to or below prey in various concentrations.

Table 2. Elemental concentration (% dry wt) of decomposer food-web constituents

Food-web constituent	N	S	Ca	Mg	K	P	Na
01 Leaves	0.94	0.22	2.24	0.18	0.19	0.07	0.007
01 Stems	0.14	0.08	2.40	0.08	0.12	0.04	0.010
02 Litter	0.96	0.16	1.50	0.14	0.14	0.07	0.009
Fungi[a]	2.80	0.40	3.30	0.19	0.12	0.24	0.040
Saprovores	7.74	0.40	10.30	0.27	0.13	0.80	0.110
Fungivores	7.74	0.40	3.95	0.46	0.40	1.39	0.600
Predators	7.74	0.40	2.50	0.32	0.18	1.06	0.370

[a]After Ausmus (1972)

Consumption of 01 litter (C_1) was estimated to be 20 $g/m^2/yr$ (Table 3). Consumption of 02 litter (C_2) was estimated to be 35 $g/m^2/yr$. Net production (E) was equivalent to 0.99 g of saprovore. Fungivores consumed 191 $g/m^2/yr$ of fungus having a net production of 30 g/m^2. Predators consumed the equivalent of 115 g of fungivore-like animals and had a net production of 29.6 g. Because predators have higher caloric equivalences, this represents nearly 35 kcal above the equivalent non-

predator mass. Since E of predators is 1/3 of C, subsequent calculations of elemental fluxes assume that 1/3 of predator diet comes
from other predators and the remainder comes from non-predators. Elemental fluxes were then calculated as the products of these dry matter
equivalents of energy flow and the elemental percentages in Table 2.

Table 3. Energy flow through a cryptozoan food-web, $kcal/m^2/yr$

Food-web Cryptozoan	C 01 Litter	C 02 Litter	ΣC	R	E
Saprovores	88.66	83.89	172.56	12.63	3.56
Fungivores	---	---	896.11	428.58	110.37
Predators	---	---	425.02	242.26	142.07

Elemental Fluxes

Since E is approximately equal to annual productivity, then the ratio
E/C is approximately equal to assimilation efficiency. For calcium,
the assimilation efficiency of 0.10 for saprovores is in exact agreement with McBrayer (1973) on which the energy calculations were based
(Table 4). Gist and Crossley (1975) showed a calcium assimilation efficiency of 0.08 for litter-feeding oribatid mites. They showed a calcium
assimilation efficiency of 0.16 for a collembolan which compares favor-
ably with the 0.19 calculated for fungivores here. Their value for
predatory mesostigmatid mites was 35%, compared to the 20% calculated
for predators here. 11.03 g calcium/m^2/yr passed through the decomposer
food-web but only 0.98 g (8.8%) was directly from litter. The remainder
entered via fungus or prey species not considered here (e.g., nematodes).
2.02 g (18%) was immobilized within invertebrate tissues.

Magnesium fluxes were about one order of magnitude less than calcium
fluxes which is in agreement with the relative concentrations in the
food-web constituents. Again, less than 10% entered directly via litter
consumption and 27% was immobilized as invertebrate production. The
calculated assimilation efficiencies are in good agreement with the
assimilation efficiencies for calcium and the relative factors by which
the elements are concentrated above substrate. For example, fungivores
contain 1.2 times as much calcium as fungi and that ratio is doubled
for magnesium. Since the elements are derived from the same food, it
follows that the assimilation efficiencies for magnesium must be twice
those for calcium.

Potassium occurs in litter in concentrations similar to magnesium, but
the ion is much more mobile. While slightly more potassium enters the
food-web via litter (13%), a similar amount is immobilized there (25%).
Best and Monk (1975) stated that leachate potassium concentrations increased by 150% while passing through the litter layer. Immobilization
by invertebrates apparently does not retain potassium in the litter
layer to any appreciable extent.

The results for sodium are somewhat anomalous. For both saprovores and
predators, about one fourth of the sodium ingested is retained in productivity while their concentration factors are about 14 and 0.7, respectively. Both assimilation values are low when compared to Van Hook
(1971). For fungivores, however, production requires 2.4 times as much
sodium as was available in the fungus eaten. Assuming the elemental
analyses to be correct, supplemental sodium would be required. In eastern soils, exchangeable sodium tends to be roughly equal to exchangeable

potassium, although plant demands are much less. Presumably, sodium could be derived from ingesting soil, in the manner of large herbivores, but no evidence for this mechanism in soil invertebrates has been demonstrated.

Table 4. Elemental fluxes (g/m^2/ yr.) through a cryptozoan food-web

Element		Saprovores	Fungivores	Predators	Sum
Ca	C	.976	6.292	3.762	11.03
	E	.101	1.178	.740	2.02
	E/C	.10	.19	.20	---
Mg	C	.085	.362	.447	.89
	E	.003	.137	.095	.24
	E/C	.03	.38	.21	---
K	C	.087	.229	.359	.68
	E	.001	.119	.053	.17
	E/C	.02	.52	.15	---
Na	C	.004	.076	.479	.56
	E	.001	.179	.110	.29
	E/C	.24	1.0+	.23	---
N	C	.525	5.338	8.211	14.07
	E	.076	2.309	2.293	4.68
	E/C	.14	.43	.28	---
P	C	.039	.458	1.377	1.87
	E	.008	.415	.314	.74
	E/C	.20	.91	.23	---
S	C	.100	.763	.424	1.29
	E	.004	.119	.118	.24
	E/C	.04	.16	.28	---

C: ingestion; E: losses ($\simeq$ productivity)

Fungi are rich in nitrogen, compared to detritus, and represent the primary source of nitrogen for the decomposer food-web. Invertebrate production represents 900% of the nitrogen in litter consumed by invertebrates and 80% of all the nitrogen entering the food-web.

Invertebrate production requires nearly 20 times as much phosphorus as enters the food-web as litter and 1.5 times as much as enters via fungi plus litter. That difference can easily be accommodated by internal cycling and conservation. Calculated assimilation efficiencies appear well-within the range of probabilities. Paris and Sikora (1967) found an assimilation efficiency of $\sim$93% for ^{32}P in isopods and ^{32}P assimilation efficiencies ranged from 50 to 65% for various developmental stages of the oribatid mite, *Damaeus clavipes* (Luxton, 1972).

Fluxes of sulfur more closely resembled those of calcium and magnesium than other anionic elements. Only 28% of incoming elemental S was required for production, and assimilation efficiencies of saprovores and fungivores were substantially lower than they were for nitrogen and phosphorus.

Discussion

In order to evaluate the impact of decomposer invertebrate production
on elemental cycles, it is necessary to relate it to the total elemental flux through the litter layer. This requires knowing litter
turnover, which is deceptively difficult to accomplish. Simply subtracting minimum litter standing crop from maximum litter standing
crop is an underestimate of decay because litter accumulates during
the season when decomposition is most rapid. Nevertheless, this
minimum litter disappearance model (hereafter the "minimum model")
predicts greater litterfall than do two ecosystem-level models of
elemental flux. The carbon model of Reichle et al. (1973) predicts
18% less litter input and the calcium model of Shugart et al. (1976)
predicts 30% less litter input. Both are based on adjacent sinkholes
with the same floral composition but lacking the significant input from
side slopes experienced in the present study. However, in the sinkholes
where these studies were based, the results of Edwards and Sollins
(1973) predict litter turnover 2.3 times as great as the minimum model.
Their estimate appears to be the best, to date.

Multiplying the Edwards and Sollins (1973) estimate of litter turnover
by the elemental concentrations determined for this study, we find
that total invertebrate production (Σ E's) accounts for only 8.1% of
the calcium in litterfall, 10% of the potassium, 13.7% of the sulfur,
and 15.6% of the magnesium. Of these, only sulfur is anionic in the
soil solution. The relatively low immobilization of sulfur is predictable from the concentration ratios in the food-chain (Table 2). Nevertheless, sulfur levels in soils were limiting in preindustrial days and
biotic retention mechanisms are expected. Decomposer invertebrates do
not appear to be significantly involved, however. For the three cationic
elements, the existence of the soil cation exchange complex obviates
evolutionary pressures on biotic retention mechanisms.
Nitrogen and phosphorus fluxes suggest the possibility of biological
conservation. Nitrogen occurs in soils in both anionic (NO_2^-, NO_3^-) and
cationic (NH_4^+ forms. While the ammonium ion is exchangeable on the
organo-clay complex, it is not thought to persist in soils. During
the growing season, ammonium is taken up directly by vegetation and
that which is not taken up is readily oxidized to anionic forms. Some
may be fixed by clay minerals but is effectively removed from the
cycle in the process (Buckman and Brady, 1969). Nitrate, the predom-
inant anion, is subject to leaching loss if not taken up by vegetation.
The productivity of decomposer invertebrates is equivalent to 70% of
the nitrogen released during litter decomposition and represents
organic immobilization of what would eventually end up as nitrate.
4.68 g/m^2 of nitrogen (Σ E) in the form of animal bodies represents
a quickly mobilized source of ammonium, with the major decline in
animal biomass coincidental with the start of the growing season. Total
invertebrate biomass declined by 3.4 g/m^2 or 0.26 g N/m^2 between January and March. But production was 8.4 g, so total nitrogen release
was 0.91 g/m^2. On the Oak Ridge Reservation, 60% of 225 plants flower
during the period from February to April (Taylor, 1969). Thus, ~14%
of the nitrogen in litterfall is being released from invertebrates as
vegetation begins growing.

The interaction with phosphorus is more dramatic. Total soil phosphorus
tends to be low. In acid soils, most phosphorus occurs as highly in-
soluble phosphates of iron or aluminum. While iron and aluminum phosphates only slowly yield phosphate ions to the soil solution, free
phosphate ions are quickly removed by them. Leaching losses of phosphate ions are much less important than removal through chemical pre-

cipitation. Phosphorus availability to vegetation becomes critical during the initiation of growth and flowering (Black, 1968). The spring invertebrate population decline would release 138 mg P/m^2, or 22% of the phosphorus returned as litter. Total annual production of invertebrates utilizes the equivalent of 118% of the phosphorus contained in litter-fall, which implies internal recycling within the food-web. This internal recycling can distribute ion release throughout the growing season.

Sodium is not abundant in eastern soils but plant requirements are low and deficiencies apparently do not exist. Animal requirements are much greater and they concentrate sodium by a factor of 15 or more above substrate. Litterfall can supply less than one third of the sodium requirements of decomposer invertebrates but the cycling problem can be solved by the ingestion of soil. The apparent problem occurs only in fungivores but they would be forced to ingest nearly as much soil as fungus to satisfy their sodium demands. The results remain anomalous and warrant further investigation.

So, are invertebrates important in regulating rates of elemental release in deciduous forest? Putting aside the question of their effect on microbial activity, it appears that invertebrates have no direct effect on release rates for essential cations. However, for nitrogen and phosphorus, there is circumstantial evidence that they are important. Nitrogen availability is supplemented by biological fixation, but that would probably be equivalent to only about 23% of invertebrate production, based upon the estimates of Todd et al. (1975). Since no comparable phosphorus input exists, invertebrate turnover of phosphorus may be especially critical.

Acknowledgments. Research was supported by the Eastern Deciduous Forest Biome Project, U.S. IBP, funded by the National Science Foundation under Interagency Agreement AG-199, 40-193-69 with the Energy Research and Development Administration-Oak Ridge National Laboratory. Contribution No. 278 from the Eastern Deciduous Forest Biome, U.S. IBP.

References

Ausmus, B.S.: Progress report on soil and litter microfloral energetics studies at the Oak Ridge site. U.S./IBP Eastern Deciduous Forest Biome, Memo Report 72-150. Oak Ridge Natl. Lab., 1972, 38 p
Best, G.R. Monk, C.D.: Cation flux in hardwood and white pine watersheds. pp. 847-861, In: Mineral Cycling in Southeastern Ecosystems. Howell, F.G., Gentry, J.B., Smith, M.H. (eds.). U.S. ERDA, CONF-740513, 1975
Black, C.A.: Soil-Plant Relationships. New York: Wiley, 1968, 792 pp
Buckman, H.O., Brady, N.C.: The Nature and Properties of Soils. New York: Macmillan, 1969, 653 pp
Edwards, N.T. Sollins, P.: Continuous measurement of carbon dioxide evolution from partitioned forest floor components. Ecology 54, 406-412 (1973)
Gist, C.S., Crossley, D.A., Jr.: 1975. A model of mineral-element cycling for an invertebrate food-web in a southeastern hardwood forest litter community. pp. 84-106, In: Mineral Cycling in Southeastern Ecosystems. Howell, F.G., Gentry, J.B., Smith, M.H. (eds.). U.S. ERDA, CONF-740513, 1975
Hook, R.I. van, Jr.: Energy and nutrient dynamics of spider and orthopteran populations in a grassland ecosystem. Ecol. Monogr. 41, 1-26 (1971)

Luxton, M.: Studies on the oribatid mites of a Danish beech wood soil.
    I. Nutritional biology. Pedobiologia 12, 434-463 (1972)
McBrayer, J.F.: Exploitation of deciduous leaf litter by *Apheloria montana*
    (Diplopoda: Eurydesmidae). Pedobiologia 13, 90-98 (1973)
McBrayer, J.F., Reichle, D.E.: Trophic structure and feeding rates of
    forest soil invertebrate populations. Oikos 22, 381-388 (1971)
McBrayer, J.F., Reichle, D.E., Witkamp, M.: Energy flow and nutrient
    cycling in a cryptozoan food-web. U.S./IBP Eastern Deciduous Forest
    Biome, EDFB-IBP-73-8. Oak Ridge Natl. Lab., 1974
Paris, O.H., Sikora, A.: Radiotracer analysis of the trophic dynamics
    of natural isopod populations. In: Secondary Productivity of Ter-
    restrial Ecosystems. Petrusewicz, K. (ed.). Warsaw: Polish Acad.
    Sci., 1967, pp. 741-771
Petrusewicz, K., Macfadyen, A.: Productivity of Terrestrial Animals,
    Principles and Methods. Oxford: Blackwell Sci. 1970, 190 pp.
Reichle, D.E.: Energy and nutrient metabolism of soil and litter in-
    vertebrates. In: Productivity of Forest Ecosystems. Duvigneaud, P.,
    (ed.). Paris: UNESCO, 1971, pp. 465-477
Reichle, D.E., Dinger, B.E., Edwards, N.T., Harris, W.F., Sollins, P.:
    Carbon flow and storage in a forest ecosystem. In: Carbon and the
    Biosphere. Woodwell, G.M., Pecan, E.V. (eds.). U.S. AEC CONF-720510,
    1973, pp. 345-365
Satchell, J.E.: Litter - Interface of animate/inanimate matter. In:
    Biology of Plant Litter Decomposition. Dickinson, C.H., Pugh, G.J.F.
    (eds.). London: Academic, 1974, pp. xiii-xliv
Shugart, H.H., Jr., Reichle, D.E., Edwards, N.T., Kercher, J.R.: A
    model of calcium-cycling in an east Tennessee *Liriodendron* forest:
    model structure, parameters and frequency response analysis. Ecology
    57, 99-109 (1976)
Taylor, F.G.: Phenological records of vascular plants at Oak Ridge,
    Tennessee. Oak Ridge Natl. Lab., 1969
Todd, R.L., Waide, J.B., Cornaby, B.W.: Significance of biological
    nitrogen fixation and denitrification in a deciduous forest eco-
    system. In: Mineral Cycling in Southeastern Ecosystems. Howell, F.G.,
    Gentry, J.B., Smith, M.H. (eds.). U.S. ERDA, CONF-740513, 1975,
    pp. 729-735

Chapter 9

Soil Microorganism—Arthropod Interactions: Fungi as Major Calcium and Sodium Sources

K. CROMACK, JR., P. SOLLINS, R. L. TODD, D. A. CROSSLEY, JR.,
W. M. FENDER, R. FOGEL, and A. W. TODD

Introduction

Biodegradation of litter components and soil organic matter is a complex process in which both microorganisms and soil animals take part. Fungi usually comprise the dominant microbial biomass in terrestrial decomposer communities, and contribute substantially to cycling of both macronutrients and trace elements (Harley, 1971; Stark, 1972; Ausmus and Witkamp, 1973). Fungi are also important as energy and nutrient sources for many vertebrates and invertebrates (Miller and Halls, 1969; Fogel and Peck, 1974; Mitchell and Parkinson, 1976). Invertebrates adapted to coprophagy can obtain increased quantities of essential elements from feces colonized by microbes (Wieser, 1966; McBrayer, 1973).

Our objective in this paper is to present evidence that terrestrial fungi may be important sources of Ca and Na for saprophagous arthropods and other soil animals. Calcium, but not Na, has been reported as essential for many fungi (Foster, 1949; Lilly, 1965); both elements are essential for animals (Prosser, 1973).

Accumulation of Calcium and Sodium by Terrestrial Fungi

Fungi accumulate substantial concentrations of macronutrients and micronutrients in a wide variety of forest ecosystems. Relative to both hardwood and coniferous forest floors, fungal rhizomorph tissues contained significantly greater concentrations of Ca, K, and Na; fungal sporocarps contained significantly greater concentrations of Cu, K, Na, and P (Cromack et al., 1975). Stark (1972, 1973) found substantial concentrations of Ca, Cu, Fe, K, Mn, N, Na, and P in fungi in a Jeffrey pine ecosystem and in tropical forests. In both types of ecosystems she found that Ca concentrations in fungal rhizomorphs and Na concentrations in both rhizomorphs and sporocarps were several-fold higher than in tree foliage (Table 1). The data indicate that both fungal hyphae and rhizomorphs generally contain Ca concentrations which are several-fold higher than their ambient substrates. Sporocarps, on the other hand, have Ca concentrations much lower than hyphae or rhizomorphs. Na concentrations are similar to those of rhizomorphs. Fogel (1976) gives data for Ca and other elements in sporocarps of hypogeous fungi.

Calcium accumulation as Ca oxalate is widespread in fungi (De Bary, 1887; Foster, 1949). This may explain in part the high Ca concentrations found in hyphae and rhizomorphs (Stark, 1972; Todd et al., 1973; Cromack et al., 1975). Oxalic acid is a low energy compound having only 8.5% the caloric value of glucose (Foster, 1949). Its common occurrence in substrates colonized by fungi may be important in soil weathering, for the oxalate anion is an extremely effective chelator of cations such as Ca, Fe, and Al (Bruckert, 1970). Formation of short chain organic anion complexes with Fe and Al may solubilize P from Fe and Al hydroxy-phosphates (Stevenson, 1967; Harley, 1975). Tree foliage also contains Ca oxalate (Chandler, 1937).

Table 1. Calcium and sodium concentrations in various organic materials

Organic material	Location	Range in elemental conc. (% Dry Wt)	
		ppm Ca	ppm Na
White pine forest floor[a]	N. Carolina	5120-7880	2-23[a,b]
Fungal hyphae[c]	" "	27,100-32,100	---
Fungal rhizomorphs[a]	" "	27,200-68,200	251-557
Fungal sporocarps[a]	" "	125-275	94-246
Jeffrey pine needles[d]	Nevada	1052	265
Fungal rhizomorphs[d]	"	6589-77,437	219-3415
Fungal fruiting bodies[d]	"	110-3625	250-900
Hardwood forest floor[a]	N. Carolina	15,620-17,220	2-28[a,b]
Fungal hyphae[c]	" "	76,900-86,300	---
Fungal rhizomorphs[a]	" "	26,800-35,200	127-157
Fungal sporocarps[a]	" "	400-1000	687-961
Hardwood forest floor[e]	Tennessee	29,000-73,000	130-450
Fungal rhizomorphs[e]	"	8100-8900	385-315
Fungal sporocarps[e]	"	1000-1200	380-420
Fungal rhizomorphs[d]	Central and	7750-111,120	616-29,125
Fungal sporocarps[d]	S. America	116-5000	210-2313

[a]Cromack et al. (1975)
[b]Yount (1975)
[c]Todd et al. (1973)
[d]Stark (1972, 1973)
[e]Ausmus and Witkamp (1973)

Calcium and Sodium in Soil Animals

Many soil animals are noted for their utilization of substantial quantities of Ca. For example, earthworms are well-known users of Ca due to calciferous glands present in certain species (Robertson, 1936). Diplopods, isopods, and snails utilize appreciable Ca in their exo-skeletons; oribatid mites also concentrate Ca (Wallwork, 1971; Cornaby, 1973; Gist and Crossley, 1975). Cornaby et al. (1975) estimated that soil fauna processed approximately 11% of Ca while ingesting 20% of litterfall biomass from the forest floor annually in a deciduous forest watershed. Oribatid mites, macroarthropods, and snails generally have substantial whole-body concentrations of Ca which they probably obtain from fungi in litter and soil (Table 2). Earthworms could utilize fungal calcium as one source of the element for their calciferous glands. Springtails, though they have low whole-body Ca concentrations, may utilize considerable Ca due to their rapid population turnover.

Sodium concentrations in all of the soil animal groups are higher than Na concentrations present in the forest floors in North Carolina and Tennessee (Table 2). Present Na data from both temperate and tropical forests suggest that Na concentrations in fungi are sufficient for them to be an important dietary source of Na for soil animals. Gosz et al. (1976) found Na residence time to be greater than that of K in the forest floor, indicating active Na concentration by heterotrophs in a hardwood forest floor. Carter and Cragg (1976) also have found Na and Ca accumulation by some species of soil arthropods in an aspen stand.

Table 2. Concentrations of calcium and sodium in soil animals

Soil animal	Location	Range in Elemental Conc. (% dry wt)	
		ppm Ca	ppm Na
Microarthropods			
Prostigmata[a]	N. Carolina	5600	5700
Prostigmata[b]	" "	300–400	---
Cryptostigmata[a]	" "	48,700–74,100	2300–6100
Cryptostigmata[b]	" "	3900–64,300	---
Cryptostigmata[c]	" "	35,000	2200
Collembola[a]	" "	1600–1700	3300–8900
Collembola[c]	" "	2957	---
Macroarthropods			
Isopoda[a]	N. Carolina	72,800–84,900	4000–9800
Isopoda[d]	Tennessee	108,900	5100
Diplopoda[a]	N. Carolina	99,500–143,100	1600–4700
Diplopoda[c]	" "	148,950	352–1282
Diplopoda[d]	Tennessee	217,500–546,300	3200–6000
Diplopoda[e]	"	419,000–462,400	3800–4100
Others			
Pulmonata[a]	N. Carolina	178,400–324,200	1900–39,800
Oligochaeta[a]	" "	4100–7200	5400–9300

[a]Cornaly (1973)
[b]Crossley (1976) Na not analyzed
[c]Gist and Crossley (1975), and Gist (personal communication)

[d]Reichle et al. (1969). Data reported as ash-free dry wt
[e]McBrayer (1973)

Calcium Cycling and Soil Animals

The decomposition of oxalate salts is of interest with regard to Ca utilization by soil animals. Fungi do not appear to break down Ca oxalate due to its low solubility. However, they can decompose soluble oxalates and oxalic acid (Foster, 1949). A number of bacteria, including *Streptomyces*, can break down Ca oxalate (Jakoby and Bhat, 1958; Chandra and Shethna, 1975). Oxalate decomposing bacteria, including *Streptomyces*, have been isolated from earthworms or their casts, indicating that these animals could obtain Ca by microbial decomposition of Ca oxalate in the gut (Jakoby and Bhat, 1958). Parle (1963) found that bacteria and actinomycetes increase in the intestines of several earthworm species. The fact that soil arthropods such as isopods can degrade hemicelluloses and aromatic compounds such as phenols, cinnamic acid, and quinic acid, possibly via specialized gut flora, make it likely that simple C compounds such as oxalate could also be decomposed in soil animal digestive systems (Reyes and Tiedje, 1976; Neuhauser and Hartenstein, 1976).

Preliminary tests for the presence of oxalate decomposers in the digestive systems of both terrestrial and aquatic detritivores were all positive (Table 3). These data provide circumstantial evidence that oxalate decomposers are a normal component of the gut flora of these animals. Decomposition of organic acid soil salts, such as Ca oxalate, in the guts of soil animals may contribute to an increase in gut pH as a moderately strong acid is converted into a much weaker one:

$$2CaC_2O_4 + O_2 \longrightarrow 2Ca^{++} + 2CO_3^{--} + 2CO_2$$

Table 3. Microbial oxalate decomposers isolated from soil animals

Soil animal	Oxalate decomposers	Reference
Oligochaeta:		
Pheretina sp.	*Pseudomonas oxalaticus*	Khambata and Bhat (1953)
Common Indian Earthworm	*Streptomyces* sp.	Khambata and Bhat (1954)
Common Indian Earthworm	*Mycobacterium lacticola*	Khambata and Bhat (1955)
Lumbricus rubellus	Bacteria and Actinomycetes	Cromack et al. (1976)
Endemic earthworm species - Oregon	Actinomycetes	Cromack et al. (1976)
Microarthropods:		
Cryptostigmata: Pelopoidea	Actinomycetes	Cromack et al. (1976)
Collembola: *Sinella* sp.	Actinomycetes	Cromack et al. (1976)
Aquatic arthropods:		
Insecta: *Peltoperla* sp.	Actinomycetes	Cromack et al. (1976)
Insecta: *Stenonema* sp.	Actinomycetes	Cromack et al. (1976)

Jayasuriya (1955) observed pH to increase from 7.0 to 9.5 during bacterial decomposition of K oxalate. Both Van Der Drift and Witkamp (1960) and McBrayer (1973) found higher pH in litter detritivore feces than in leaf litter prior to digestion; results which could have been due in part to organic acid salt decomposition. Increase in soil animal gut pH might increase their tolerance of polyphenolic and humic acid anions during digestion of litter.

A proposed Ca cycle operative in fungi, bacteria, and soil animals in the context of the soil ecosystem, is presented in Figure 1. In this diagram Ca is depicted as cycling either within the soil animal digestive system or externally within the soil ecosystem. As depicted in Figure 1, Ca may exist in several forms: as Ca on exchange sites in soil or litter, as Ca bicarbonate, as Ca oxalate, or as Ca^{++} in solution. In this simplified diagram, we omit comprehensive detail of other Ca compounds existing in soil or litter or within the soil animal. It is quite possible that the proposed system could operate in aquatic detrivore systems as well. A point worth emphasizing in the context of the proposed cycle is that waste products of the C cycle may profoundly influence cycling of elements such as Ca, and in a more general context, influence cycling of elements such as P, Fe, Al, and others.

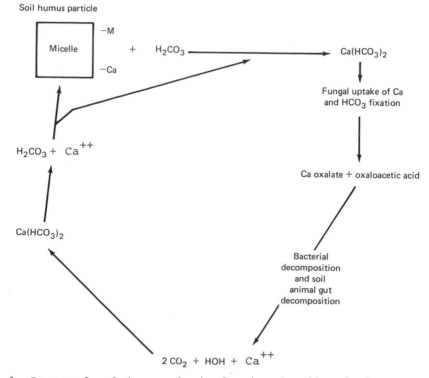

Fig. 1. Proposed calcium cycle in fungi and soil animals

Further studies are needed on microflora in soil animal digestive systems such as the one by Parle (1963) to confirm a resident oxalate decomposer flora. Radioisotope tagging, using ^{45}Ca and ^{14}C-labeled oxalic acid would be useful in confirming oxalate decomposition in digestive systems of a variety of soil animals.

Conclusions

It is concluded that terrestrial fungi are important potential sources of Ca and Na and other elements for saprophagous arthropods and other soil animals. The extent soil fauna, which show considerable evolutionary divergence, appear to share the characteristic of substantial Ca utilization. Insects, which do not appear to concentrate Ca to the same extent as other saprophagous arthropods, may process substantial quantities of the element in ecosystems where they have rapid population turnover. Both mycorrhizal and non-mycorrhizal fungi appear to share common nutrient cycling mechanisms in some instances, though their physiology differs in other respects.

Acknowledgments. This research was supported by NSF grants GB-20963 and AG-199, 40-193-69 and supplement #147 to the master Memorandum of Understanding between the USDA Forest Service and Oregon State University. This is contribution No. 264 from the Coniferous Forest Biome and No. 272 from the Eastern Deciduous Forest Biome. The use of the

Central Chemical Laboratory facilities with the assistance of Elly Holcombe and the use of the Electron Microscope Laboratory facilities with the assistance of R.B. Addison at the U.S. Forestry Sciences Laboratory, Pacific Northwest Forest and Range Experiment Station, Project PN-1653, is acknowledged. This is contribution No. 1105 from the Forest Research Laboratory, Oregon State Univ.

References

Ausmus, B.S., Witkamp, M.: Litter and soil microbial dynamics in a deciduous forest stand. USAEC Rep. No. EDFB-IBP 73-10, Oak Ridge Nat. Lab., 1973

de Bary, A.: Comparative Morphology and Biology of the Fungi, Mycetozoa and Bacteria. Oxford: Clarendon (Engl. transl.), 1887

Bruckert, S.: Influence des composés organiques solubles sur la pédogenèse en milieu acide. I. Etudes en terrain. Ann. Agron. 21, 421-452 (1970)

Carter, A., Cragg, J.B.: Concentrations and standing crops of calcium, magnesium, potassium, and sodium in soil and litter arthropods and their food in an aspen woodland ecosystem in the Rocky Mountains (Canada). Pedobiologia 16, 379-388 (1976)

Chandler, R.F.: Certain relationships between the calcium and oxalate content of foliage of certain forest trees. J. Agr. Res. 55, 393-396 (1937)

Chandra, T.S., Shethna, Y.I.: Isolation and characterization of some new oxalate-decomposing bacteria. Anton von Leeuwen. 41, 101-111 (1975)

Cornaby, B.W.: Population parameters and systems models of litter fauna in a white pine ecosystem. Ph.D. dissertation, Univ. Georgia, 1973

Cornaby, B.W., Gist, C.S., Crossley, D.A., Jr.: Resource partitioning in leaf litter faunas from hardwood and hardwood-converted-to-pine forests. In: Mineral Cycling in Southeastern Ecosystems. Howell, F.G., Gentry, J.B., Smith, M.H. (eds.). Springfield, VA: Technical Information Service, ERDA, 1975, pp. 588-597

Cromack, K., Jr., Todd, R.L., Monk, C.D.: Patterns of basidiomycete nutrient accumulation in conifer and deciduous forest litter. Soil Biol. Biochem. 7, 265-268 (1975)

Cromack, K, Jr., Sollins, P., Todd, R.L., Fogel, R., Todd, A.W., Fender, W.M., Crossley, M.E., Crossley, D.A., Jr.: The role of oxalic acid and bicarbonate in calcium cycling by fungi and bacteria: Some possible implications for soil animals. In: Soil Organisms as Components of Ecosystems. Lohm, U., Persson, T. (eds.). Ecol. Bull. (Stockholm) Vol. 25 (in press), 1977

Crossley, D.A., Jr.: Oribatid mites and nutrient cycling. In: Proc. Symp. on Soil Mites (1976, in press)

Van Der Drift, J., Witkamp, M.: The significance of the breakdown of oak litter by *Enoicyla pusilla*. Burm. Arch. Neerland. Zool. 13, 486-492 (1960)

Fogel, R.: Ecological studies of hypogeous fungi. II. Sporocarp phenology in a western Oregon Douglas-fir stand. Can. J. Botany 54, 1152-1162 (1976)

Fogel, R., Peck, S.B.: Ecological studies of hypogeous fungi. I. Coleoptera associated with sporocarps. Mycologia 67, 741-747 (1974)

Foster, J.W.: Chemical Activities of Fungi. New York: Academic, 1949

Gist, C.S., Crossley, D.A., Jr.: The litter arthropod community in a southern Appalachian hardwood forest: Numbers, biomass and mineral element content. Am. Midl. Naturalist 93, 107-121 (1975)

Gosz, J.R., Likens, G.E., Bormann, F.H.: Organic matter and nutrient dynamics of the forest and forest floor in the Hubbard Brook Forest. Oecologia 22, 305-320 (1976)

Harley, J.L.: 1971. Fungi in ecosystems. J. Ecol. 59, 653-668 (1971)

Harley, J.L.: Problems of mycotrophy. In: Endomycorrhizas. Mose, B., Tinker, P.B. (eds.). New York: Academic, 1975, pp. 1-24

Jakoby, W.B., Bhat, J.V.: Microbial metabolism of oxalic acid. Bateriol. Rev. 22, 75-80 (1958)

Jayasuriya, G.C.N: The isolation and characteristics of an oxalate-decomposing organism. J. Gen. Microbiol. 12, 419-428 (1955)

Khambata, S.R., Bhat, J.V.: Decomposition of oxalate by *Streptomyces*. Nature (London) 174, 696-697 (1954)

Khambata, S.R., Bhat, J.V.: Decomposition of oxalate by *Mycobacterium lacticola* isolated from the intestine of earthworms. J. Bacteriol. 69, 227-228 (1955)

Khambata, S.R., Bhat, J.V.: Studies on a new oxalate-decomposing bacterium, *Pseudomonas oxalaticus*. J. Bacteriol. 66, 505-507 (1953)

Lilly, V.G.: Chemical constituents of the fungal cell. In: The Fungi. Ainsworth, G.F., Sussman, A.S. (eds.). New York: Academic, 1965, Vol I, pp. 163-177

McBrayer, J.F.: Exploitation of deciduous leaf litter by *Apheloria montana* (Diplopoda: Eurydesmidae). Pedobiologia 13, 90-98 (1973)

Miller, H.A., Halls, L.K.: Fleshy fungi commonly eaten by southern wildlife. USDA, Forest Service, Southern Forest Exptl. Sta. Res. Paper SO-49, 1969

Mitchell, M. W., Parkinson, D.: Fungal feeding of oribatid mites (Acari: Cryptostigmata) in an aspen woodland soil. Ecology 57, 302-312 (1976)

Neuhauser, E., Hartenstein, R.: Degradation of phenol, cinnamic and quinic acid in the terrestrial crustacean, *Oniscus asellus*. Soil Biol. Biochem. 8, 95-98 (1976)

Parle, J.N.: Micro-organisms in the intestines of earthworms. J. Gen. Microbiol. 31, 1-11 (1963)

Prosser, C.L.: Comparative Animal Physiology. Philadelphia: Saunders, 1973, Vol I

Reichle, D., Shanks, M.H., Crossley, D.A., Jr.: Calcium, potassium and sodium content of forest floor arthropods. Ann. Ent. Soc. Am. 62, 57-62 (1969)

Reyes, V.G., Tiedje, J.M.: Metabolism of ^{14}C-labelled plant materials by woodlice *(Tracheoniscus rathkei* Brandt) and soil microorganisms. Soil Biol. Biochem. 8, 103-108 (1976)

Robertson, J.D.: The function of the calciferous glands of earthworms. J. Exptl. Biol. 13, 279-297 (1936)

Stark, N.: Nutrient cycling pathways and litter fungi. Bioscience 22, 355-360 (1972)

Stark, N.: Nutrient Cycling in a Jeffrey Pine Ecosystem. Missoula: Univ. Montana Press, 1973

Stevenson, F.J.: Organic acids in soil. In: Soil Biochemistry. McLaren, A.D., Peterson, G.H. (eds.). New York: Marcel Dekker, 1967, pp. 119-146

Todd, R.L., Cromack, K., Jr., Stormer, J.C., Jr.: Chemical exploration of the microhabitat by electron probe microanalysis of decomposer organisms. Nature (London) 243, 544-546 (1973)

Wallwork, J.A.: Some aspects of the energetics of soil mites. In: Proc. Third Intern. Cong. of Acarology. Daniel, M., Rosicky, B. (eds.). The Hague: Junk, 1971, pp. 129-134

Wieser, W.: Copper and the role of isopods in degradation of organic matter. Science 153, 67-69 (1966)

Yount, J.D.: Forest floor and nutrient dynamics in southern Appalachian hardwood and white pine plantation ecosystems. In: Mineral Cycling in Southeastern Ecosystems. Howell, F.G., Gentry, J.B., Smith, M.H. (eds.). Springfield: Technical Information Service, ERDA, 1975, pp. 598-608

Chapter 10

Ant Nests as Accelerators of Succession in Paraguayan Pastures

J. C. M. JONKMAN

Introduction

Certain pasture areas in the South American Gran Chaco convert rapidly
into woodlands. Nests of the leaf cutting ant, *Atta vollenweideri*, which
occur in these pastures, have an accelerating effect on succession.
In addition, certain land use partices seem to cause increased nest
densities and expanded distribution of ants.

The genus *Atta* occurs from the southern United States to the north
of Argentina (Weber, 1972). *Atta vollenweideri* is the southernmost representative of the 14 recognized species (Borgmeier, 1959). It occurs
throughout the entire Gran Chaco-region in northern Argentina and
Western Paraguay. The occurrence of *A. vollenweideri* in this vast region
shows two distinct distribution areas, one being a fairly narrow strip
along the Paraguay River, and the other area corresponding to the entire remaining part of the Chaco. The principal difference between
these two subregions is that in the Paraguay River subregion, *A. vollenweideri* occurs only in pastures. In the remaining subregion it occurs
in bushland and under the thorny Chaco scrubs; but also in agricultural
clearings (Jonkman, 1976).

Methods

By means of aerial photographs I inspected approximately 80,000 km^2
of the Paraguay River subregion for the occurrence of ant nests. For
this survey I used photographs at scale 1:50,000 and 1:60,000 dating
from 1965 and 1968, respectively. In addition, I examined 1968 photographs at scale 1:10,000 for more detailed interpretation, individual
trees being distinguishable at this scale. Moreover, vertical photographs at scale 1:38,000 and oblique photographs were available, dating
from 1944. The adult ant nests are clearly visible on the small scale
aerial photographs as white points, whereas heavy trees and clusters
of shrubs are also distinguishable.

Ant Nests

The ant nests are visible even on the small scale photographs because
of their enormous size and the contrast between the nest mound, consisting of very light-colored soil, and the darker surrounding grass.
The average surface area of a nest mound is approximately 40 m^2. Because of erosion of the mound, an even larger area may be distinguished
as the "nest site," which has an average surface of approximately 75
m^2.

An adult nest consists of a few thousand cavities of which the majority are "living chambers" containing a sponge-like substrate. The substrate is principally made of grass particles and serves as a growth base for fungi, the ants' food supply. Old substrate residues are deposited in 50 to 100 large refuse chambers, which occur at the bottom of the nest (up to a depth of almost 5 m).

The soil, which is excavated in the construction of the chambers, is deposited on top of the mound and has a higher pH (up to 8.5) than the surrounding soil (5.7 and higher). As a consequence of this pH difference, the eroded zone around the nest remains bare or is covered by only short grasses. This condition may continue for many years after the nest no longer contains a living colony.

Plant Invasion of Nest Sites

The development of vegetation on ant nests usually begins after the death of the inhabiting colony. Some time after a colony dies, its nest collapses because the protection mechanisms constructed by the ants against penetration of water are no longer maintained. The collapse might also be accelerated by two species of *Posopodidae* (armadillos), which preferably make their dens in dead ant nests by digging rather deep and broad tunnels. The collapsed center of the nest often contains a little pool of water. Afterwards the bare nest site is slowly invaded by several plant species: grasses begin to grow on the edge of the water-containing center and cactus species (especially *Opuntia* spp.) are among the first to settle around the elevated parts of the nest.

In the next phase various shrubs appear, together with young Caranday palms, *Copernicia* spp., and trees, mostly Leguminosae: *Prosopis campestris* (espinillo), *Prosopis alba* (algarrobo blanco), *Prosopis algarrobilla* (algarrobillo). Sometimes the shrubs are only sparcely present. As the vegetation grows higher, the pool in the center of the old nest dries up. Eventually a "wood nucleus" forms on top of the old nest (Fig. 1). However, a clear border is maintained between the white-colored outer zone around the nest, and the surrounding grass. At this stage the former nest structure is still recognizable. The species composition of the wood nucleus, is different from that of the nearby forests. In fact, the *Prosopis* species never occur there.

An extreme example of wood nuclei development occurred in a pasture which had 2.3 nests/ha in 1944. Over a 29-year period (1944-1973) woody vegetation invaded all the nests and thereby changed the character of the pasture environment.

The invasion and development of vegetation on old ant nests resembles vegetation development on termite mounds (Troll, 1936; MacFayden, 1950; Glover et al., 1964). Thomas (1941) explained that islands of woody vegetation begin on termite mounds and then gradually increase in size so that eventually the "islands" of adjacent mounds merge with one another.

The impact of *A. vollenweideri* has not been fully assessed, but is related to its distribution and abundance. The ant occurs in a few "main areas" of the Paraguay River subregion. These main areas constitute about 10% of the 80,000 km² area which I inspected. Only 13% of the main areas were occupied which translates to roughly 1.3% of the inspected area.

Fig. 1. Typical "wood nucleus" along with opuntias on an overgrown nest site. The two Caranday palms probably developed before the ant nest. Most woody vegetation in the background consists of other "wood nuclei"

In such occupied areas, only 1% of the ground surface was covered by nest sites since nest densities averaged about 1.3/ha.

Acknowledgments. I thank Prof. Dr. K. Bakker, Dr. B. de Crombrugghe, and C.H. van Vierssen for their comments on the manuscript. This study was made possible by the United Nations Educational, Scientific, and Cultural Organization (UNESCO), the Dutch Government, and the Institute of Basic Sciences in Asuncion, Paraguay.

References

Borgmeier, T.: Revision der Gattung Atta. Fabr. (Hym. Formicidae). Studia Entomol. 2, 321-390 (1959)
Glover, P.E., Trump, E.C., Wateridge, L.E.D.: Termitaria and vegetation patterns on the Loita Plains of Kenya. J. Ecol. 52, 367-377 (1964)
Jonkman, J.C.M.: Biology and ecology of the leaf cutting ant Atta vollenweideri Forel, 1893. Z. Ang. Ent. 81, 140-148 (1976)
MacFayden, W.A.: Vegetation patterns in the semi desert plains of British Somaliland. Geogr. J. 116, 199-211 (1950)
Thomas, A.S.: The vegetation of the Sese Islands. Uganda J. Ecol. 29, 330-353 (1941)
Troll, C.: Termiten Savannen. In: Festschrift Norbert Krebs, Stuttgart: Englehorn, 1936
Weber, N.A.: Gardening Ants, the attines. Philadelphia: Am. Philosoph. Soc., 1972, pp. 23-26

Chapter 11

Community Structure of Collembola Affected by Fire Frequency

D. L. Dindal and L. J. Metz

Introduction

The Collembola are among the most common insects found in forest soils.
They are an important segment of the soil mesofauna and contribute to
decomposition processes by reducing organic material to smaller bits,
thereby increasing the surface area making it more susceptible to attack by fungi and bacteria. They also feed on and spread fungal spores
and bacteria (Christen, 1975; Chap. 9 this vol.). By these actions
they contribute to the cycling of nutrients in the forest.

This paper reports a study of the effects of prescribed burning on one
facet of the community structure of Collembola, namely the presence
and magnitude of interspecific associations.

Location and Conditions of Study

Prescribed burning is commonly used in southern pine forests of the
United States to reduce fire hazard, kill young hardwood trees, and
to prepare seedbeds for regeneration of pine. This study was installed
within loblolly pine, *Pinus taeda*, stands on the Santee Experimental
Forest in Berkeley County, South Carolina, about 25 years ago to in-
vestigate the effects of various types of burns on the vegetation and
soils. Data reported in this paper were collected over a period of ten
months on three different kinds of plots: those not burned (controls),
those burned every four to six years (periodic burns), and those burned
every year (annual burns). This treatment scheme provided an ideal
chance to observe fire as a selection pressure and to note both positive and negative responses exhibited by biota. More specific information on the study site and initial experimental methods can be found
in Metz and Farrier (1973).

The normal forest floor of a loblolly pine stand consists of three
layers: the surface litter (L) layer made up of recently fallen material which is relatively undecayed; the fermentation (F) layer which
is immediately below the L and shows evidence of decay as the material
is darkened and in smaller pieces; and the humus (H) layer which has
undergone decomposition and is a black amorphous material unrecogniz-
able as to origin.

Although 100 separate samples were collected from each layer of the
forest floor and from each of the surface 3 cm of mineral soil, the
community structure analyses discussed here were confined to the F
and H layers, because the vast majority (about 86%) of collembolans
occurred there. Finally, about 1500 individual collembolans representing 78 species were extracted and identified for the study. The 25
most common of these species along with a corresponding alphabetical
symbol are presented in Table 1. A species was arbitrarily selected
as "common" if it was found to occur in $\geq$ 5 associations.

Table 1. Dominant collembolan species encountered in study on prescribed burning

Species	Symbol
Folsomina onychiurina Denis	G
Harlomillsia oculata (Mills)	I
Hypogastrura armata Nicolet	J
Isotoma trispinata Macillivray	KA
Isotomiella minor Schaeffer	M
Lepidocyrtus cyaneus var. *cinereus* Folsom	P
Lepidocyrtus cyaneus var. *assimilis* Reuter	O
Lepidocyrtus curvicollis Bourlet	Q
Megalothorax incertoides Mills	R
Neanura barberi Handschin	SA
Neanura muscorum Templeton	SB
Neosminthurus curvisetis (Guthrie)	U
Orchesella ainsliei Folsom	XA
Orchesella villosa Geoffrey	XD
Proisotoma besselsi Packard	YA
Proisotoma constricta Folsom	YE
Proisotoma minuta Tullberg	Y
Pseudosinella violenta Folsom	DD
Sminthurinus elegans Fitch	GG
Tomocerus flavescens var. *separatus* Folsom	JJ
Tullbergia clavata Mills	LL
Tullbergia granulata Mills	MM
Tullbergia iowensis Mills	MMA
Xenylla grisea Axelson	OO
Xenylla welchi Folsom	PP

The presence of interspecific associations was evaluated using the chi-square methods of Cole (1949). Magnitudes of the associations were calculated using Cole's coefficient. This coefficient gives the degree of association as well as the positive or negative nature of the species' interactions. This paramenter was determined only for associations that were significant at $Pr \leq .10$.

Results and Discussion

Community Diversity

From basic structural analyses of the communities we reported (Metz and Dindal, 1975) the following general patterns: (1) species diversity increased after prescribed burning (both periodic and annual); (2) species richness was more important than equitability in the diversity measures; and (3) the frequency of dominance and the dominant species were generally reduced by prescribed burns. As diversity and dominance fluctuated in relation to fire, so did the presence and magnitude of significant relationships between species.

Presence of Significant Interactions

Control Sites

On control sites (Fig. 1), there were significant interspecific associations (probability (Pr) <10%) exhibited by 17 species. These species were related in 25 different ways with three associations in the
≤ 1% probability class, 12 associations in the 2-5% group and ten associations in the 6-10% Pr classes. Of special note for comparative pur-
poses, is the highly significant relationship of species *Lepidocyrtus cyaneus* and *Tullbergia clavata* (O-LL). Also, the moderately significant level
of the *Hypogastrura armata* to *Folsomina onychiurina* (J-G) combination in the
control community is important.

Periodic Burn Sites

Seventeen species exhibiting 21 different associations were found in the
periodically burned sites (Fig. 2). The distribution of the probability
classes are: eight associations, Pr ≤ 1%; six associations, Pr 2-5%;
seven associations, Pr 6-10%. Although the total number of associations
at the ≤ 10% level has been reduced by four, the highly significant
relationships were increased from three on the control to eight on the
periodic burn site. The genera, *Lepidocyrtus* and *Tullbergia* were again represented by the highly significant pairing of *Lepidocyrtus curvicollis*
and *Tullbergia iowensis* (Q-MMA). The moderately significant relationship of
L. cyaneus and *Neosminthurus curvisetis* (O-U), not found in the control, probably developed in response to the burn. Also, moderate interactions of
L. cyaneus to *Tullbergia granulata* (O-MM) appeared only in the periodic burn
sites, not in either of the others. Lussenhop (1976) found these same
genera present on biennially burned prairie communities in Wisconsin.
No species interactions were given; however, he reported that *T. iowensis*
increased significantly (Pr 5-10%) in response to burning.

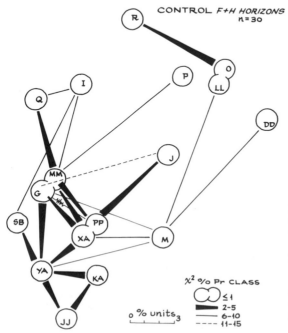

Fig. 1. Constellation diagram of interspecific associations of Collembola—prescribed burning in loblolly
pine stands. "% units" scale
represents subdivisions within each probability class
except in relationships where
graphic foreshortening could
not be avoided. For explanation see text

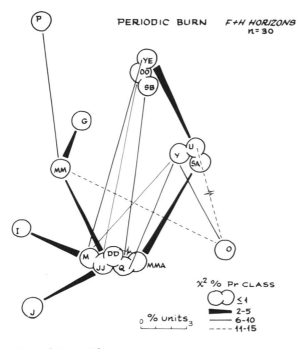

Fig. 2. Constellation diagram of interspecific associations of Collembola—prescribed burning in loblolly pine stands. "% units" scale represents subdivisions within each probability class except in relationships where graphic foreshortening could not be avoided. For explanation see text

Annual Burn Sites

The severe stress of annual burning (Fig. 3) further reduced the number of species involved in significant interactions to only seven. Four such interactions occurred at Pr levels < 10%, two associations at Pr 2-5% and two associations at Pr < 1%. These represent in a minor way a colonizing nucleus capable of maintaing a degree of community organization.

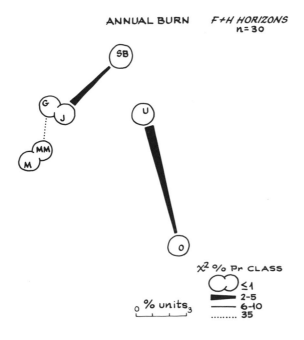

Fig. 3. Constellation diagram of interspecific associations of Collembola—prescribed burning in loblolly pine stands. "% units" scale represents subdivisions within each probability class except in relationships where graphic foreshortening could not be avoided

Although representatives of *Tullbergia* (MM) and *Lepidocyrtus* (O) were present, the "association dominance" noted among these genera on other sites was not present here. Additional differences exist between significant species combinations mentioned as being found on other sites. For example, when compared to the control, the species association *H. armata* and *F. onychiurina* (J-G) has been "congealed" from a moderate to a highly significant one. Also, the *L. cyaneus assimilis* to *N. curvisetis* (O-U) relationship became more significant after the fire treatment, whereas, the species relationship *F. onychiurina* and *T. granulata* (G-MM) has been gradually disrupted from the control by the fire factor.

Association Frequency

Figure 4 illustrates the results of considering fire as a selection factor that affected interspecific relationships. Species (represented by alphabetical symbols) are plotted versus percentage of other species with which each species is related, thus providing an "association frequency." Using this scheme, five "selection-association response" categories were evident. The "selection-association response" category and number of species characterizing each are as follows: *reduced but stable* (4 spp); *fire-eurysensitive*, showing some degree of burn tolerance (3 spp); *stenosensitive to fire*, showing no burn tolerance (7 spp); *fire-tolerant*, affected very little or even increased by fire (3 spp); and *association pioneers* or *colonizers*, relationships apparently stimulated and only found on post-burn sites (8 spp). Of special note among "association colonizers" are *Sminthurus elegans* (GG) and *Orchesella villosa* (XD) both of which were not related with any other species on any sites except under the maximum stress.

Magnitude of Interspecific Associations

In general, increased fire frequency resulted in fewer negative species assocations. Thirty-six percent of the associations observed in the control site were negative. Perhaps these can be equated with negative feedback processes which may promote community stability. Such processes as competition or various other interspecific relationships that have a negative component as outlined by Dindal (1975) could be involved. On periodically burned sites the frequency of the negative associations dropped to 20%, and on annual burn sites there were no negative relationships.

The relationships of Cole's coefficient within all sites are reviewed in Figure 5; a perfect positive or negative correlation is equal to 1. Very similar configurations of the positive associations were observed between the control and periodic burn data, whereas, the magnitude of positive interactions on the annual area were quite weak. Furthermore, the dynamics of selected species associations (Fig. 6) shows that fire does change the magnitude of interrelationships. Species-specific association patterns exist and appear to exhibit plasticity dependent upon the burn condition. Therefore, a certain response occurs in magnitude as well as presence of association when collembolans are subjected to fire as a selection pressure.

Summary

The genera *Lepidocyrtus* and *Tullbergia* are represented on all sites and are generaly associated with each other or with burned site conditions. Per-

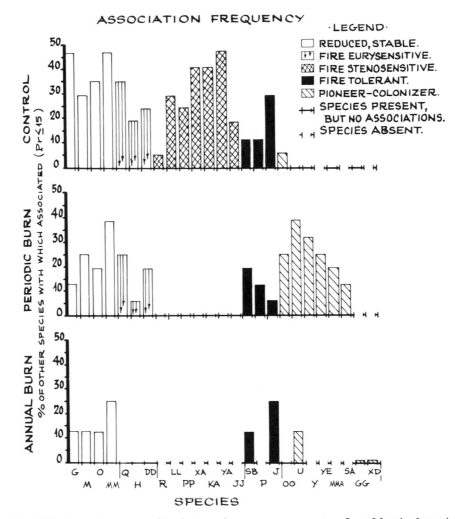

ASSOCIATION FREQUENCY

·LEGEND·

☐ REDUCED, STABLE.
⊞ FIRE EURYSENSITIVE.
▨ FIRE STENOSENSITIVE.
■ FIRE TOLERANT.
⧄ PIONEER-COLONIZER.
⊢⊣ SPECIES PRESENT, BUT NO ASSOCIATIONS.
⊣ ⊦ SPECIES ABSENT.

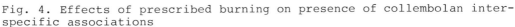

Fig. 4. Effects of prescribed burning on presence of collembolan inter-specific associations

haps the differences in species that are found in these two genera are indicative of similar niches being filled under slightly different microhabitat conditions.

Not only are individual species sensitive or tolerant to the action of fire, but also interspecific associations exhibit like sensitivities and tolerances. Comparing responses from the control to the most stressed site, there are reductions in the frequency of negative interspecific associations ranging from 36% to 0. Although the numerical complexity and number of associations are reduced by periodic burning some semblance of order within the community is retained. However, major reductions, as seen on annual burn sites, could have dramatic effects on negative feedback loops and community stability. Therefore, it appears that fire can cause shifts for and against certain species and associations of Collembola, thus modifying their total community structure.

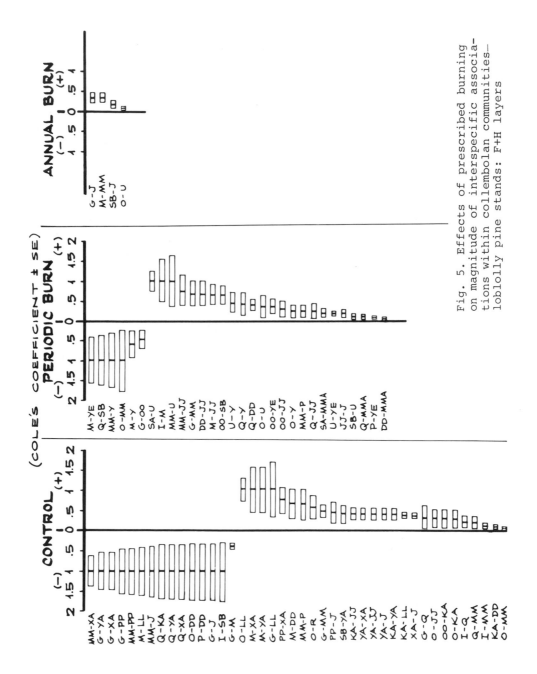

Fig. 5. Effects of prescribed burning on magnitude of interspecific associations within collembolan communities—loblolly pine stands: F+H layers

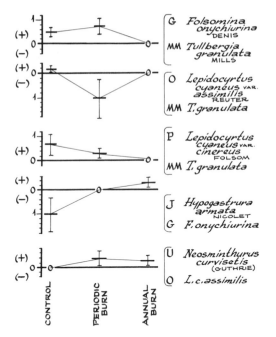

Fig. 6. Effects of prescribed burning on magnitude of selected associations of collembolans-loblolly pine stands: F+H layers

Acknowledgment. We thank Dr. and Mrs. David Wray for assistance in identifying the collembolan specimens.

References

Christen, A.A.: Some fungi associated with Collembola. Rev. Ecol. Biol. Sci. 12, 723-728 (1975)

Cole, L.C.: The measurement of interspecific association. Ecology 30, 411-424 (1949)

Dindal, D.L.: Symbiosis: Nomenclature and proposed classification. Biologist 57, 129-142 (1975)

Lussenhop, J.: Soil arthropod response to prairie burning. Ecology 57, 88-98 (1976)

Metz, L.J., Dindal, D.L.: Collembola populations and prescribed burning. Environ. Entomol. 4, 583-587 (1975)

Metz, L.J., Farrier, M.H.: Prescribed burning and populations of soil mesofauna. Environ. Entomol. 2, 433-440 (1973)

Chapter 12

Saprophagous Organisms and Problems in Applied Resource Partitioning

B. W. CORNABY

Introduction

The allocation or partitioning of resources by species or groups of species is measured by how they utilize energy, matter, space, and time.
These interacting resources form gradients in ecosystems and numerous
scientists are exploring the gradients singly and in combination relative
to their use by organisms (Levins, 1968; Roughgarden, 1974; Southwood
et al., 1974; Wiens, 1976). Examples of resource partitioning by saprophagous arthropods are numerous. Schoener (1974) reviewed studies of
termites (Sands, 1965), niche segregation in seven millipede species
(O'Neill, 1967), and vertical distributions of mite species (Hulbert,
1968). Some other studies of resource partitioning by saprophages include
those inhabiting animal carcasses (Cornaby, 1974) and dung (Rainio, 1966).
Cornaby et al. (1975) reported different rates of calcium and potassium
utilization by communities of leaf-litter fauna in contrasting forests.

The bulk of the literature on the soil/litter subsystem and its fauna
is discipline-oriented and emphasizes the interplay and synthesis of
theory and observation from an evolutionary time-frame. Furthermore,
many of the theories and models have been generated from the perspective
of an unmodified, pristine world. However, in reality, man-modified
nature is the rule (Detwyler, 1971; Jacobs, 1975), and many current
problems require solutions in an ecological time-frame.

This paper presents three resource problems that command worldwide attention: (1) recognizing levels of environmental quality, (2) accelerating the natural regeneration of spent land, and (3) deactivating path-
ologic materials and decomposing compounds that are not easily biode-
graded. Such problems are now being defined and addressed. The solutions
of each problem require the interplay and synthesis of theory, observation, and application. Saprophagous animals have a part in this synthesis, and one of the needed end products is an improved understanding
of the role of arthropods in forests and other ecosystems.

Biological Indicators

The need for good biological indicators of environmental quality has
accelerated since the implementation of the U.S. National Environmental
Policy Act (NEPA Public Law 91-190), subsequent guidelines from the U.S.
Council on Environmental Quality, and similar laws in other countries.
An indicator is something (species or complex of species) that reveals
important aspects of structure and function for some portion of the ecosystem without exhaustive study of that part. Good indicator data can
be synthesized to permit a quantitative ranking of habitats in terms of
their relationship to certain standards.

Saprophagous as well as phytophagous arthropods can be utilized as indicators of environmental quality (Cornaby, 1975; Duke, 1975). Table 1
presents one approach to using phytophagous arthropods as indicators
in an environment dominated by mixed upland forest in West Virginia,
U.S.A. (Battelle, 1975). Species diversity (Shannon-Weiner) measure-

ments of the arthropod communities were computed for sweep-samples from several sites. Locations with diversities of less than 4.0 suffered from minor disturbances, e.g., adjacent dirt roads and movement of vehicles. On the other hand, sites such as the cove, slopes, and ridges which were distant from minor disturbances showed higher diversities. Thus, the arthropod diversities seemed to track the disturbance level of the vegetation. Indicator measurements for birds, small mammals, reptiles, vegetation, soil, and water corroborated trends indicated by the arthropods. Together, the indicator measurements provide insight into the nature of the resource gradients which, in turn, are interpretable from the perspective of environmental quality.

Table 1. Relationships of diversity indices for samples of arthropod communities to various patches of vegetation in the R.D. Bailey, West Virginia, U.S.A. area

Approximate location	Species diversity	Disturbance level of vegetation-patch
Floodplain 5	5.43	least disturbed
Cove 1	5.39	
South-facing slope 10	5.16	
Ridge 3	5.15	
North-facing slope 7	4.92	
Ridge 4	4.83	
South-facing slope 9	4.63	
Argus Hollow	4.47	
Pigeon Creek	4.29	
Guyandotte River	4.03	
Indian Creek	3.77	intermediate
Pigeon Creek	3.42	
Guyandotte River	3.34	
Logging Road	3.00	most disturbed

The use of terrestrial arthropods and other organisms as indicators is in its infancy. A sound theoretical base for such ecoassays and better observational methods are needed to identify the best indicators and to interpret the observations relative to environmental quality. One of the most promising approaches is the use of resource gradients under two unifying themes: cover (primarily space and time) and food (primarily energy and matter). These two key elements of carrying capacity can provide a quantitative basis for evaluating and assessing the quality of resources from the ecological perspective. Also, such concepts have a commonality with economic and social aspects of environmental quality.

Land Regeneration

Improved utilization of saprophagous arthropods and other soil organisms is essentail to the ultimate regeneration of spent land. Stripped land, resulting from only one of the major activities that alters the soil/plant/animal systems, totals 1,782,000 hectares in the United States alone (Grim and Hill, 1974). It is increasing at the rate of 81,000 hectares per year and about 50% of the total is associated with coal mining. These lands will require various management inputs to restore them to an acceptable steady state condition. Since soil arthropods have a significant role in soil genesis, their part in integrated, ecosystems approaches to reclaiming land must not be overlooked.

Prior to stripmining or other vegetation-removing practices, forest floors exhibited certain mosaics. These mosaics of physiochemical conditions normally change through space and time. Temperature, moisture, light, soil porosity, and diverse organic materials, are some of the variables which contribute to substrate variety. This diversity or

98

heterogeneity can be viewed as a series or gradient of patches. Levin and Paine (1974) referred to a patch as a hole or a discontinuity. Organisms key their activities to the environmental discontinuity. For example, if an individual organism or species spends its entire life in a single patch, the environments are coarse-grained. If an organism needs resources from many patches, the environment is fine-gained for that particular species (Levins, 1968). The overall number, permanence, and quality of coarse- and fine-grained environments will determine the distribution and abundance of cover and food, which, in turn, dictate how populations of saprophagous organism will interact. For example, it is predicted that organisms can take food preferentially in coarse-grained environments, but must take food in the same proportions in which it occurs in fine-grained environments (MacArthur and Pianka, 1966). Additional insight into the importance of habitat mosaics may be found in Wiens (1976).

Disruption of land changes the shapes, sizes, and inter-patch distances and this results in new kinds of mosaic. A certain amount of disturbance of ecological succession creates and maintains spatiotemporal heterogeneity, and Levin and Paine (1974) have developed a mathematical model to predict some of these relationships. However, severe alterations can result in mosaics that are disadvantageous to some and advantageous to other soil fauna. Field studies of millipedes (Neuman, 1973) and Collembolla (Bode, 1975) have shown that these and other soil organisms occur in certain densities and species compositions on soil banks (man-created patches). Vimmerstedt and Finney (1973) conducted greenhouse tests with earthworms in spoil-bank soil and predicted that earthworms would be useful in incorporating organic matter into mineral soil under field conditions. When such organism-oriented studies are integrated into habitat-oriented approaches, better predictive models can be developed for the directed use of saprophagous arthropods, annelids and other organisms. Knowledge of the changes in spatial and temporal heterogeneity is a key to helping solve the problem.

Deactivation of Noxious and Decomposition of Slowly Degrading Materials

Potentially harmful substances are being introduced at increasing rates into ecosystems. The possible role of saprophagous arthropods and other soil organisms in concentrating and deactivating polycyclic organic matter, carcinogens, and other noxious materials awaits investigation. For 300 substances reported as being carcinogenic, mutagenic, or teratogenic (McCann et al., 1975), there is considerable potential for the use of saprophagous animals as intervening agents in the flow of pathological substances from sources to man.

Yet another problem is the slow biodegradation of high molecular weight hydrocarbon polymers. The biodegradability of such synthetic substances is still questionable according to Heap and Morrell (1968) and Mills and Eggins (1970). However, a systematic screening of enzyme systems among the soil/litter organisms may reveal some which have potential for degrading h-m-w hydrocarbon polymers.

These kinds of problems can be approached through the use of the theories of feeding strategies. It is the intent of this theory to understand how an organism's behavior and morphology are organized to locate and process food (energy and matter) (Schoener, 1971). Potential foods include noxious and high molecular weight material. The theory should allow us to predict if an organism's ecological feeding strategy will

accommodate carcinogenic and high-molecular weight polymers. Four
questions become important. What is the optimal diet; what is the optimal size of the feeding patch (is it part of a coarse- or fine-grained
environment?); what is the optimal feeding time; and what is the optimal
feeding-group size? Another aspect of this theory relates to the organism's ability to metabolically assimilate certain kinds of materials.
Additional insight will be gained from the use of such information as
ratios of beneficial (calcium, zinc) to metabolically analogous, and
potentially harmful materials (strontium, cadmium) in organisms and
food chains (Lemons and Kennington, 1976). Thus, if certain species
serve as intervening, ameliorating agents or accelerate biodegradation,
we could develop breeding programs for such soil/litter organisms. In
turn, these complements of organisms could be innoculated into strategi-
cally located soil/litter systems whose assimilative capacities were
overloaded. Ultimately, the challenge becomes one of how to use an
organism's genetic fitness or optimization (Schoener, 1971; Cody, 1974)
to solve ecological problems. Surely, solutions to problems in deacti-
vation and biodegradation will emerge as a synthesis of feeding
strategies, nutrient cycling, and related concepts.

Conclusion

I have dramatized three real-world problems in applied resource partitioning. Biological indicators, land regeneration, and particularly de-
activation of noxious substances by saprophagous organisms stimulate
intellectual interest. I have related theories about carrying capacity,
patchy environments, and feeding strategies to these problems and suggested that future research must be a synthesis of theory, observation,
and application. Indeed, each intellectual activity can mutually benefit
when developed in concert. It is hoped that the problems in this paper
will attract the serious attention of other scientists and scholars and
that their solutions will provide another dimension to our knowledge of
saprophagous arthropods in forest ecosystems.

Acknowledgments. I thank Dr. K.M. Duke and Dr. J.T. McGinnis for review of
the manuscript and the ecosystem/engineering group at Battelle for
creating an atmosphere wherein theory is being correlated with application and observation.

References

Battelle Columbus Laboratories: R.D. Bailey Lake Project, Final Report
    Environmental Assessment to Dept. Army, Huntington District, Corps
    of Engineers, Huntington, West Virginia, 1975, 244 p
Bode, E.: Pedozoologische Sukzessionsuntersuchungen auf Rekultivierungs-
    flächen des Braunkohlentagebaues. Pedobiologia 15, 284-289 (1975)
Cody, M.L.: Optimization in ecology. Science 183, 1156-1164 (1974)
Cornaby, B.W.: Carrion reduction by animals in contrasting tropical
    habitats. Biotropica 6, 51-63 (1974)
Cornaby, B.W.: Soil arthropods as indicators of environmental quality.
    In: Organisms and Biological Communities as Indicators of Environ-
    mental Quality-a Symposium. King, C.C., Elfner, L.E. (eds.). Ohio
    Biol. Surv. Inform. Circ. 8, 23-25 (1975)
Cornaby, B.W., Gist, C.S., Crossley, D.A., Jr.: Resource partitioning

in leaf-litter faunas from hardwood and hardwood-converted-to-pine forests. In: Mineral Cycling in Southeastern Ecosystems, Howell, F.G., Gentry, J.B., Smith, M.H. (eds.). ERDA Symp. Ser. (CONF-740513), 1975, pp. 588-597

Detwyler, T.R.: Man's Impact on Environment. New York: McGraw-Hill, 1971, 731 p

Duke, K.M.: Above-ground arthropods as indicators of biological quality. In: Organisms and Biological Communities as Indicators of Environmental Quality—a Symposium. King, C.C., Elfner, L.E. (eds.). Ohio Biol. Surv. Inform. Circ. 8, 25-26 (1975)

Grim, E.C., Hill, R.D.: Environmental protection in surface mining of coal. U.S. EPA 670/2-74-093, Industrial Waste Treatment Res. Lab. Cincinnati, Ohio, 1974, 277 p

Heap, W.M., Morrell, S.H.: Microbiological deterioration of rubber and plastics. J. Appl. Chem. 18, 189-194 (1968)

Hurlbutt, H.W.: Coexistence and anatomical similarity in two genera of mites, *Veigaia* and *Asca*. Syst. Zool. 17, 261-271 (1968)

Jacobs, J.: Diversity, stability and maturity in ecosystems influenced by human activities. In: Unifying Concepts in Ecology. Van Dobben, W.H., Lowe-McConnell, R.H.: (eds.). The Hague: Junk, 1975, pp. 187-207

Lemons, J.D., Kennington, G.S.: Distributions and natural levels of related metals in a trophic pathway. Idaho Natl. Engineering Lab., ICP-1095, Idaho Falls, Idaho, 1976, 50 p

Levin, S.A., Paine, R.T.: Disturbance, patch formation, and community structure. Proc. Natl. Acad. Sci. 71, 2744-2747 (1974)

Levins, R.: 1968. Evolution in Changing Environments. Princeton, New Jersey: Princeton Univ., 1968, 120 p

MacArthur, R.H., Pianka, E.R.: On optional use of a patchy environment. Am. Naturalist 100, 603-609 (1966)

McCann, J., Choi, E., Yamasaki, E., Ames, B.N.: Detection of carcinogenesis as mutagens in *Salmonella* microsome test: Assay of 300 chemicals. Proc. Natl. Acad. Sci, 72, 5135-5139 (1975)

Mills, J., Eggins, H.O.W.: Growth of thermophilic fungi on oxidation products of polyethylene. Intern. Biodeterior. Bull. 6, 13-17 (1970)

Neumann, U.: Succession of soil fauna in afforested spoil banks of the brown-coal mining district of Cologne. In: Ecology and Reclamation of Devastated Land. Hutnik, R.J., David, G. (eds.). New York: Gordon and Breach, 1973, Vol. I, pp. 335-348

O'Neill, R.V.: Niche segregation in seven species of diplopods. Ecology 48, 983 (1967)

Rainio, M.: Abundance and pheonology of some coprophagous beetles in different kinds of dung. Ann. Zool. Fennici 3, 88-98 (1966)

Roughgarden, J.: Population dynamics in a spatially varying environment: how population size "tracks" spatial variation in carrying capacity. Am. Naturalist 108, 649-664 (1974)

Sands, W.A.: Termite distribution in man-modified habitat in West Africa, with special reference to species segregation in the genus *Trinervitermes* (Isoptera, Termitidae, Nasutitermitinae). J. Animal Ecol. 34, 557-572 (1965)

Schoener, T.W.: Theory of feeding strategies. Ann. Rev. Ecol. Syst. 2, 369-404 (1971)

Schoener, T.W.: Resource partitioning in ecological communities. Science 185, 27-39 (1974)

Southwood, T.R.E., May, R.M., Hassell, M.P., Conway, G.R.: Ecological strategies and population parameters. Am. Naturalist 108, 791-804 (1974)

Vimmerstedt, J.P., Finney, J.H.: Earthworm introduction on litter burial and nutrient distribution in Ohio strip-mine spoil banks. Soil Sci. Soc. Am. Proc. 37, 338-391 (1973)

Wiens, J.A.: Population responses to patchy environments. Ann. Rev. Ecol. Sys. 7, 81-120 (1976)

Subject Index

Modern Methods in Forest Genetics

Edited by J.P. Miksche

With Contributions by
F. Bergmann, G.P. Berlyn, W. Bücking, R.A. Cecich, L.S. Dochinger,
F.H. Evers, P.P. Feret, R.B. Hall, K.M. Hansen, O. Huhtinen, K.F. Jensen,
E.G. Kirby, J. Lunderstädt, J.P. Miksche, R.P. Pharis, B.R. Roberts,
A.E. Squillace, R.G. Stanley, A.M. Townsend, R.B. Walker,
K. von Weissenberg, L. Winton, W. Zelawski

1976. xiv, 288p. 38 illus.
Proceedings in Life Sciences
ISBN 0-387-07708-1

Contents
Optical Techniques for Measuring DNA Quantity—Nucleic Acid Extraction,
Purification, Reannealing, and Hybridization Methods—Gel Electrophoresis of
Proteins and Enzymes—Extraction and Analysis of Free and Protein-Bound Amino
Acids from Norway Spruce Foliage—Photosynthesis, Respiration, and Dry Matter
Production—Analyses of Monoterpenes of Conifers by Gas-Liquid Chromatography—
Isolation and Analysis of Plant Phenolics from Foliage in Relation to Species
Characterization and to Resistance Against Insects and Pathogens—Mineral
Analyses—Pollution Responses—Indirect Selection for Improvement of Desired
Traits—Pollen Handling Techniques in Forest Genetics, with Special Reference to
Incompatibility—Tissue Culture of Trees—Manipulation of Flowering in Conifers
Through the Use of Plant Hormones.

Contributions by specialists in their respective fields cover modern techniques
applicable to forest genetics and breeding. Among the subjects discussed, and where
possible demonstrated in laboratory procedures, are microspectro-photometric
measurement and biochemical characterization of nucleic acids, amino acids, proteins
and enzymes, and their extraction and assay using electrophoresis and thin layer
chromatographs in the perspective of the interpretation of genetic variation
of relatedness.

Further chapters treat photosynthetic efficiency in the production of carbohydrate
polymers and the determination of resin derivatives and phenolics as related to
inheritance, genetic considerations, with protocols for mineral analysis, and the
impact of air pollution on forest trees. Also presented are indirect selection,
pollination techniques, protoplast fusion, and hormone flowering manipulation
procedures. This carefully edited book is not only for specialists, but also for
lecturers and advanced students.

BIOGEOCHEMISTRY
of a Forested Ecosystem

By G.E. Likens, F.H. Bormann, R.S. Pierce,
J.S. Eaton, and N.M. Johnson

1977. xii, 146p. 37 illus.
ISBN 0-387-90225-2

Contents

Based on the well-known "Hubbard Brook" ecosystem studies, *Biogeochemistry of a
Forested Ecosystem* presents the most in-depth analysis of the Biogeochemistry of any
terrestrial ecosystem. It brings together long-term data on precipitation and
stream-water chemistry, hydrology, and weathering and also considers the dynamics
of atmospheric gases and water as they flow through the system. The book illustrates
ways in which the ecosystem is affected by the three major biogeochemical vectors of
the earth: air, water, and organisms. In turn, it shows how the ecosystem moderates
and changes inputs and how it affects biogeochemical cycles by its outputs. Acid
precipitation is an important example of the ways in which inadvertent human
activities influence atmospheric inputs in remote areas. The book illustrates that
ecosystem control over biogeochemical functions is highly predictable and relatively
repeatable from year to year. The original data from the "Hubbard Brook" studies
are compared with data from diverse ecosystems throughout the world.